Springer Complexity

Springer Complexity is an interdisciplinary program publishing the best research and academic-level teaching on both fundamental and applied aspects of complex systems – cutting across all traditional disciplines of the natural and life sciences, engineering, economics, medicine, neuroscience, social and computer science.

Complex Systems are systems that comprise many interacting parts with the ability to generate a new quality of macroscopic collective behavior the manifestations of which are the spontaneous formation of distinctive temporal, spatial or functional structures. Models of such systems can be successfully mapped onto quite diverse "real-life" situations like the climate, the coherent emission of light from lasers, chemical reaction-diffusion systems, biological cellular networks, the dynamics of stock markets and of the internet, earthquake statistics and prediction, freeway traffic, the human brain, or the formation of opinions in social systems, to name just some of the popular applications.

Although their scope and methodologies overlap somewhat, one can distinguish the following main concepts and tools: self-organization, nonlinear dynamics, synergetics, turbulence, dynamical systems, catastrophes, instabilities, stochastic processes, chaos, graphs and networks, cellular automata, adaptive systems, genetic algorithms and computational intelligence.

The three major book publication platforms of the Springer Complexity program are the monograph series "Understanding Complex Systems" focusing on the various applications of complexity, the "Springer Series in Synergetics", which is devoted to the quantitative theoretical and methodological foundations, and the "SpringerBriefs in Complexity" which are concise and topical working reports, case-studies, surveys, essays and lecture notes of relevance to the field. In addition to the books in these two core series, the program also incorporates individual titles ranging from textbooks to major reference works.

Understanding Complex Systems

Founding Editor: S. Kelso

Future scientific and technological developments in many fields will necessarily depend upon coming to grips with complex systems. Such systems are complex in both their composition – typically many different kinds of components interacting simultaneously and nonlinearly with each other and their environments on multiple levels – and in the rich diversity of behavior of which they are capable.

The Springer Series in Understanding Complex Systems series (UCS) promotes new strategies and paradigms for understanding and realizing applications of complex systems research in a wide variety of fields and endeavors. UCS is explicitly transdisciplinary. It has three main goals: First, to elaborate the concepts, methods and tools of complex systems at all levels of description and in all scientific fields, especially newly emerging areas within the life, social, behavioral, economic, neuro- and cognitive sciences (and derivatives thereof); second, to encourage novel applications of these ideas in various fields of engineering and computation such as robotics, nano-technology and informatics; third, to provide a single forum within which commonalities and differences in the workings of complex systems may be discerned, hence leading to deeper insight and understanding.

UCS will publish monographs, lecture notes and selected edited contributions aimed at communicating new findings to a large multidisciplinary audience.

More information about this series at http://www.springer.com/series/5394

Antonios Garas

Editor

Interconnected Networks

 Springer

Editor
Antonios Garas
Chair of Systems Design
ETH Zürich
Zürich, Switzerland

ISSN 1860-0832   ISSN 1860-0840   (electronic)
Understanding Complex Systems
ISBN 978-3-319-79560-7   ISBN 978-3-319-23947-7   (eBook)
DOI 10.1007/978-3-319-23947-7

Springer Cham Heidelberg New York Dordrecht London
© Springer International Publishing Switzerland 2016
Softcover reprint of the hardcover 1st edition 2016

Printed on acid-free paper

Springer International Publishing AG Switzerland is part of Springer Science+Business Media (www.springer.com)

Preface

Complex networks are powerful allies of our quest to tackle complexity in all of science. Many lines can be written about the benefits of using networks to study complex systems. Nevertheless, if I had to name their single most appealing property, I would say *simplicity*. One can map the interacting elements of any system to a set of nodes, and connect these nodes with a set of links according to their interactions. That is all it takes to build a network. Such a powerful abstraction allows to study many seemingly unrelated systems with a unified set of tools, and allows different scientific fields to benefit from advancements in other disciplines. Thus, it comes as no surprise that network science keeps growing in popularity.

However, while most early results about networks and their properties were obtained under the assumption that networks are isolated, in reality many networks interact with other networks. Consider, for example, our modern societies where individuals participate in different online social networks while they maintain a sizable amount of off-line contacts. These individuals are the means by which different social networks interact, so that information can propagate from one network to the others. Other examples include, but are not limited to, technological or infrastructure networks, whose proper function may depend on the function of another network, or transportation networks which are usually organized in layers that provide complementary access to different locations. In order to understand this bigger picture, it became clear recently that we have to extend our complex networks framework, and we are now able to treat such interconnected systems as *multilayered* networks.

In a multilayered representation each individual layer represents an isolated network from the set of networks that describe the whole system, as shown in Fig. 1. The presence of links between different networks (layers) can alter the way an interconnected system of networks behaves, even though this interconnectivity does not alter the basic characteristics of the individual networks in terms of function and topology (e.g., a communication network remains a communication network even though it is connected to the power grid). Note that the multilayered view is not just another way to describe communities in a single network, as it allows to describe systems with different types of interactions among and within the various

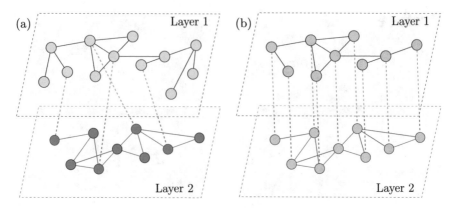

Fig. 1 Example of a two-layered network with different interconnectivity patterns. (**a**) The general case of a two-layered interconnected network where the two layers can have different number of nodes and not all nodes in one layer are interconnected to nodes of the other layer. (**b**) The special case of a multiplex network, where the nodes in the two layers are exactly the same. In this case there is a one-to-one connection between the nodes of both layers to represent their identity relation. Note, however, that in both cases, the links within the layers can be different, as indicated here with *red* and *black lines*

layers. Indeed, this nontrivial coupling allows for nonlinear effects and feedback loops, which generate emergent features that are visible only through the system as a whole and disappear when studying its individual component networks. Therefore, understanding the role of the links that connect individual layers (*interconnecting links*) and the way their presence affects the behavior of interconnected networks is a crucial step toward a more accurate description of real systems.

However, because of the different ways individual layers can be interconnected, and because interconnecting links may have different functions with respect to normal interlayer links, different naming schemes appeared aiming to distinguish cases of interest. But, instead of increasing clarity, names such as *interconnected networks*, *networks of networks*, *interdependent networks*, *multiplex networks*, etc. dominated the literature creating confusion about their actual meaning and proper use, especially with respect to what is different between them.

In order to clarify this subject, throughout this book we will call *interconnected* multilayered (or multilayer) networks the general case where there is no particular assumption with respect to the connectivity patterns and/or the function of the *interconnecting links*. Networks with such general connectivity structures are also called *interacting networks* or *networks of networks* in the literature. The special cases where the same set of nodes appear across different layers while the links within the layers are different are called *multiplex networks*. Multiplex networks are useful to describe different categorical relationships between nodes, like a set of people communicating via different channels (like phone calls, emails, etc.).

In the general case interconnecting links provide the means of interaction between networks. But if the functional properties of these links induce dependency relations, so that nodes from one layer depend on nodes form other layers to function properly, then the system of networks is called *interdependent network*. Such networks are very important, especially when studying critical infrastructures and systemic risk, because nodes that may seem safe from the single network perspective may have exposures via dependency links to other networks, which make them extremely fragile. Of course, various combinations of network-to-network connectivities with different functional properties of interconnecting links are allowed. Thus, one may encounter multiplex networks with dependency links, or networks with partial dependency links where one layer depends on another and not vice versa, etc.

With this volume we want to provide a collection of works that highlight and summarize recent developments on network theory, signaling the emergence of a mathematical, computational, and algorithmic framework that deals with interconnected complex networks, both on a theoretical and practical level. Individual chapters deal with related but in most cases complementary subjects. Each chapter is self-contained and can stand on its own. For the interested reader this removes the need to follow a particular order and allows to focus on specific subjects. However, the structure of the book follows, indeed, a specific pattern, starting with the more theoretical works and gradually dealing with more practical subjects and applications.

More precisely, the book consists of the following chapters:

- Chapter 1 shows that the formation of interconnected networks undergoes a structurally sharp (discontinuous) transition, depending on the relative importance of the links within and across layers.
- Chapter 2 describes the topology of an interconnected system of networks in terms of matrices and discusses about several metrics that are key to characterize multilayer networks and their spectral properties.
- Chapter 3 investigates diffusion dynamics on multilayer networks when we have incomplete knowledge about the link formations inside or across the layers, using ensembles of interconnected networks with similar characteristics.
- Chapter 4 describes how choosing the adequate connector links between networks may promote or hinder different structural and dynamical properties of a particular network.
- Chapter 5 provides a review of recent advances on the role of connectivity and dependency links in the robustness of interconnected networks, focusing on the dynamics of cascading failures on interdependent networks.
- Chapter 6 uses percolation theory to describe damage resilience of interconnected (multiplex) networks, following two alternative definitions on the pruning process that alter the nature of the percolation transition.
- Chapter 7 explores how much interconnectivity is needed for the emergence of cooperation in interconnected networks and shows that an intermediate density of sufficiently strong interactions between networks is the optimal case.

- Chapter 8 analyzes the influence of a time delay on a system of two interconnected networks of oscillators and explores its dynamics as a function of the couplings and communication lag.
- Chapter 9 deals with the architecture of real urban mobility networks from the multiplex network's perspective using empirical data of mobility patterns in two cities. This reveals that the socioeconomic characteristics of the population have an extraordinary impact on the layer organization of these systems.
- Chapter 10 provides a new understanding of the social structure of elites by analyzing the community structure of the generalized K-core and by identifying weakly connected regions that bridge core communities on a multiplex system, using data from a Massive Multiplayer Online Game.
- Chapter 11 reviews the empirical structure of the multiplex interbank networks and the theoretical consequences of this representation using Maximum Entropy null models.
- Chapter 12 describes the phenomenology of multilevel financial networks by reviewing selected theoretical and empirical works providing arguments in favor of adopting the broad view of the network approach to finance.

Closing this short introduction, I would like to thank all authors for their contributions and for their fruitful collaboration. Even though there are much more to be discussed about interconnected multilayered networks than what is covered in this volume, I do believe that the reader will find this collection both inspiring and motivating. I would also like to thank Frank Schweitzer for his valuable guidance that made this book possible and to acknowledge support from the EU FET project MULTIPLEX 317532.

Zürich, Switzerland   Antonios Garas
2015

Contents

Contributors

Jacobo Aguirre Centro Nacional de Biotecnología (CSIC), Madrid, Spain

José S. Andrade Jr. Computational Physics, IfB, ETH Zürich, Zürich, Switzerland

Departamento de Física, Universidade Federal do Ceará, Ceará, Brazil

Nuno A.M. Araújo Departamento de Física, Faculdade de Ciências, Universidade de Lisboa, Lisboa, Portugal

Centro de Física Teórica e Computacional, Universidade de Lisboa, Lisboa, Portugal

Alex Arenas Departament d'Enginyeria Informàtica i Matemàtiques, Universitat Rovira i Virgili, Tarragona, Spain

Leonardo Bargigli Dipartimento di Scienze per l'Economia e l'Impresa, Universitá di Firenze, Firenze, Italy

Amir Bashan Channing Division of Network Medicine, Brigham Women's Hospital and Harvard Medical School, Boston, MA, USA

Stefano Battiston Department of Banking and Finance, University of Zürich, Zürich, Switzerland

Gareth J. Baxter Department of Physics & I3N, University of Aveiro, Aveiro, Portugal

Yehiel Berezin Department of Physics, Bar Ilan University, Ramat Gan, Israel

Javier M. Buldú Laboratory of Biological Networks, Center for Biomedical Technology, UPM, Madrid, Spain

Complex Systems Group, Universidad Rey Juan Carlos, Madrid, Spain

Guido Caldarelli IMT Alti Studi Lucca, Lucca, Italy

ISC-CNR, Rome, Italy

LIMS, London, UK

Alessio Cardillo Laboratoire de Biophysique Statistique, École Polytechnique Fédérale de Lausanne (EPFL), Lausanne, Switzerland

Instituto de Biocomputación y Física de Sistemas Complejos, Universidad de Zaragoza, Zaragoza, Spain

Davide Cellai MACSI, Department of Mathematics and Statistics, University of Limerick, Limerick, Ireland

Bernat Corominas-Murtra Section for Science of Complex Systems, Medical University of Vienna, Vienna, Austria

Emanuele Cozzo Institute for Biocomputation and Physics of Complex Systems (BIFI), Zaragoza, Spain

Department of Theoretical Physics, University of Zaragoza, Zaragoza, Spain

Marco D'Errico Department of Banking and Finance, University of Zürich, Zürich, Switzerland

Michael M. Danziger Department of Physics, Bar Ilan University, Ramat Gan, Israel

Giovanni di Iasio Directorate General For Economics, Statistics and Research, Bank of Italy, Rome, Italy

Sergey N. Dorogovtsev Department of Physics & I3N, University of Aveiro, Aveiro, Portugal

A. F. Ioffe Physico-Technical Institute, St. Petersburg, Russia

Guilherme Ferraz de Arruda Departamento de Matemática Aplicada e Estatística, Instituto de Ciências Matemáticas e de Computação, Universidade de São Paulo – Campus de São Carlos, São Carlos, SP, Brazil

Antonios Garas Chair of Systems Design, ETH Zürich, Zürich, Switzerland

Jesús Gómez-Gardeñes Departamento de Física de la Materia Condensada, Universidad de Zaragoza, Zaragoza, Spain

Instituto de Biocomputación y Física de Sistemas Complejos, Universidad de Zaragoza, Zaragoza, Spain

Alexander V. Goltsev Department of Physics & I3N, University of Aveiro, Aveiro, Portugal

A. F. Ioffe Physico-Technical Institute, St. Petersburg, Russia

Ricardo Gutiérrez Department of Chemical Physics, The Weizmann Institute of Science, Rehovot, Israel

Shlomo Havlin Department of Physics, Bar Ilan University, Ramat Gan, Israel

Hans J. Herrmann Computational Physics, IfB, ETH Zürich, Zürich, Switzerland

Departamento de Física, Universidade Federal do Ceará, Ceará, Brazil

Rafael Hurtado Departamento de Física, Universidad Nacional de Colombia, Bogotá, Colombia

Luigi Infante Directorate General For Economics, Statistics and Research, Bank of Italy, Rome, Italy

Fabrizio Lillo Scuola Normale Superiore, Pisa, Italy

Laura Lotero Departamento de Ciencias de la Computación y de la Decisión, Universidad Nacional de Colombia, Medellín, Colombia

Vitor H.P. Louzada Computational Physics, IfB, ETH Zürich, Zürich, Switzerland

José F.F. Mendes Department of Physics & I3N, University of Aveiro, Aveiro, Portugal

Yamir Moreno Institute for Biocomputation and Physics of Complex Systems (BIFI), Zaragoza, Spain

Department of Theoretical Physics, University of Zaragoza, Zaragoza, Spain

Complex Networks and Systems Lagrange Lab, Institute for Scientific Interchange, Turin, Italy

David Papo Laboratory of Biological Networks, Center for Biomedical Technology, UPM, Madrid, Spain

Matjaž Perc Faculty of Natural Sciences and Mathematics, University of Maribor, Maribor, Slovenia

Federico Pierobon European Central Bank, Frankfurt am Main, Germany

Filippo Radicchi Center for Complex Networks and Systems Research, School of Informatics and Computing, Indiana University, Bloomington, IN, USA

Francisco A. Rodrigues Departamento de Matemática Aplicada e Estatística, Instituto de Ciências Matemáticas e de Computação, Universidade de São Paulo – Campus de São Carlos, São Carlos, SP, Brazil

Ricardo Sevilla-Escoboza Centro Universitario de los Lagos, Universidad de Guadalajara, Enrique Díaz de Leon, Paseos de la Montaña, Lagos de Moreno, Jalisco, México

Louis M. Shekhtman Department of Physics, Bar Ilan University, Ramat Gan, Israel

Ingo Scholtes Chair of Systems Design, ETH Zürich, Zürich, Switzerland

Frank Schweitzer Chair of Systems Design, ETH Zürich, Zürich, Switzerland

Attila Szolnoki Institute of Technical Physics and Materials Science, Research Centre for Natural Sciences, Hungarian Academy of Sciences, Budapest, Hungary

Stefan Thurner Section for Science of Complex Systems, Medical University of Vienna, Vienna, Austria

Santa Fe Institute, Santa Fe, NM, USA

IIASA, Schlossplatz 1, Laxenburg, Austria

Zhen Wang Department of Physics Hong Kong Baptist University, Kowloon Tong, Hong Kong

Center for Nonlinear Studies and the Beijing-Hong Kong-Singapore Joint Center for Nonlinear and Complex Systems, Hong Kong Baptist University, Kowloon Tong, Hong Kong

Nicolas Wider Chair of Systems Design, ETH Zürich, Zürich, Switzerland

Chapter 1
A Tipping Point in the Structural Formation of Interconnected Networks

Alex Arenas and Filippo Radicchi

Abstract The interaction substrate of many natural and synthetic systems is well represented by a complex mesh of networks where information, people and energy flows. These networks are interconnected with each other, and present structural and dynamical features different from those observed in isolated networks. While examples of such dissimilar properties are becoming more abundant, for example diffusion, robustness and competition, it is not yet clear where these differences are rooted. Here we show that the composition of independent networks into an interconnected network of networks undergoes a structurally sharp transition, a tipping point, as the interconnections are formed. Depending on the relative importance of inter- and intra- layer connections, we find that the entire interconnected system can be tuned between two regimes: in one regime, the various layers are structurally decoupled and they act essentially as independent entities; in the other regime, strong structural correlation arise, and network layers are indistinguishable i.e. the whole system behaves as a single-level network. We analytically show that the transition between the two regimes is discontinuous even for finite size networks. Thus, any real-world interconnected system is potentially at risk of abrupt changes in its structure, which may manifest new dynamical properties.

1.1 Introduction

The fundamental goals of network science are: to describe the structure of interactions between the components, and to assess the emergent behavior of many-body systems coupled to the underlying structure. Advances on the theory of complex networks will improve our understanding and modeling capabilities so that we

A. Arenas (✉)
Departament d'Enginyeria Informàtica i Matemàtiques, Universitat Rovira i Virgili, 43007 Tarragona, Spain
e-mail: alex.arenas@urv.cat

F. Radicchi
Center for Complex Networks and Systems Research, School of Informatics and Computing, Indiana University, Bloomington, IN, USA
e-mail: f.radicchi@gmail.com

© Springer International Publishing Switzerland 2016
A. Garas (ed.), *Interconnected Networks*, Understanding Complex Systems,
DOI 10.1007/978-3-319-23947-7_1

1

may control or predict the dynamics and function of complex networked systems. In addition, this approach does not rely on a detailed knowledge of the systems components and therefore allows universal results to be obtained that can be generalized with relative ease (e.g., the study of epidemic spreading processes is equivalent to the spread of computer viruses). For example, biological networks like protein interaction networks share many structural (scale-freeness) and dynamical (functional modules) features with other seemingly different systems such as the Internet and interaction patterns in social systems. Thus, systems as diverse as peer-to-peer networks, neural systems, socio-technical phenomena or complex biological networks can be studied within a general unified theoretical and computational framework.

However, almost all of the work to date is based on an ordinary 1-layer or simplex view of the networks in question, where every edge (link) is of the same type and consequently considered at the same temporal and topological scale. Generally speaking, the description of networks so far has been developed using a snapshot of the connectivity, this connectivity being a reflection of instantaneous interactions or accumulated interactions in a certain time window. This description is limiting when trying to understand the intricate variability of real complex systems, which contain many different time scales and coexisting structural patterns forming the real network of interactions. These more realistic *multi-layer* structures have received a lot of attention from the physicist community [17, 38] with no common terminology yet.

Interacting, interdependent or multiplex networks are different ways of naming the same class of complex systems where networks are not considered as isolated entities but interact with each other. In multiplex, the nodes at each network are instances of the same entity, thus the networks are representing simply different categorical relationships between entities, and usually categories are represented by layers. Interdependent networks is a more general framework where nodes can be different at each network.

Many, if not all, real networks are "coupled" with other real networks. Examples can be found in several domains: social networks (e.g., Facebook, Twitter, etc.) are coupled because they share the same actors [60]; multimodal transportation networks are composed of different layers (e.g., bus, subway, etc.) that share the same locations [4, 18]; the functioning of communication and power grid systems depend one on the other [10]. So far, all phenomena that have been studied on interdependent networks, including percolation [10, 58], epidemics [31, 32, 55], and linear dynamical systems [30], have provided results that differ much from those valid in the case of isolated complex networks. Sometimes the difference is radical: for example, while isolated scale-free networks are robust against failures of their nodes or edges [2], scale-free interdependent networks are instead very fragile [10, 58].

The standard approach towards the characterization of topological and dynamical properties of multiplex networks is similar to the one used for isolated networks. This approach relies on a fundamental approximation about the local structure of the network, generally indicated as tree-like approximation [1, 20, 21, 46]. The tree ansatz assumes the absence of finite loops in a network in the thermodynamic limit and the presence of only infinite loops. Such an approximation is very convenient because it allows one to use techniques typical of the theory of random branching processes [34]. These mainly include degree-based mean field calculations, and the application of the generating function formalism for the statistical characterization of structural and dynamical properties of ensembles of networks [48, 63]. Under this ansatz, the solutions of many problems, that are unsolvable in their exact form, can be instead provided with very good accuracy [20]: percolation [15], epidemiological [51] and opinion dynamical models [59, 61], controllability [42] are just among the most celebrated examples. The same type of approach has been applied to predict the behavior of special types of critical phenomena in multilayered networks. Examples include the analysis of the nature of the percolation transition in multiplex networks [5, 10, 28, 57, 58] and interconnected networks [27, 36, 50], and the study of the features of several dynamical processes defined on these particular type of network topologies [9, 19, 55].

Another theoretical approach used in the characterization of networks is the one based on the analysis of the spectrum of special operators associated with the graphs. This approach often relies on analytic results obtained in the branch of mathematics research known as "Spectral Graph Theory" [13], and it has been proved to be effective in the study of topological and dynamical properties of networks. Fundamental features of networks can be understood by looking at the eigenvalues and eigenvectors not only of the adjacency matrix, but also of other matrices associated with the graph such as the normalized [13] and combinatorial [45] laplacians, the non-backtracking matrix [35, 41], the modularity matrix [47], just to mention a few of them.

The fundamental reason behind the effectiveness of spectral methods is that the spectrum of a graph encodes fundamental physical features of the system: eigenvalues correspond to energy levels, and the corresponding eigenvectors represent configurations of the system associated with them. Spectral graph theory has a wide range of applicability. For example, many useful measures, such as graph energy [12], graph conductance and resistance [22], and the Randić index [39], are quantifiable in terms of the eigenvalues of the normalized laplacian of a graph. Finding the minimal eigenpair of matrices associated with graphs is typically equivalent to identifying the ground state of wide class of energy functions [40] and fitness landscapes [54]. Examples include, among others, Ising spin models [44] and combinatorial optimization problems such as the traveling salesman problem [33]. In the study of isolated networks, the spectrum of the matrices associated with graphs has been successfully applied to several contexts: examples include percolation transition [8], synchronization [3] and epidemiological models [11, 29].

Spectral approaches have recently been proved successful also for the understanding of structural [53] and dynamical [30, 52, 56] properties of networks of networks. In the study of coupled networks, the great advantage of spectral methods, with respect to those based on the tree-like approximation and the use of the generating function formalism, is in their ability to predict the behavior of multiplexes on the basis of features of the individual layers that composed them.

In the remaining part of the chapter, we will illustrate a concrete example of a successful application of spectral techniques to the characterization of structural transitions in arbitrary multiplex networks.

1.2 Mathematical Modeling

Multiplex Networks Composed of Two Layers

For simplicity, we first consider the case of two interconnected networks. We will later generalize the method to an arbitrary number of interconnected networks. We assume that the two interconnected networks A and B are undirected and weighted, and that they have the same number of nodes N. The weighted adjacency matrices of the two graphs are indicated as A and B, respectively, and they have both dimensions $N \times N$. With this notation, the element $A_{ij} = A_{ji}$ is equal to the weight of the connection between the nodes i and j in network A. The definition of B is analogous.

We consider the case of one-to-one symmetric interconnectivity [10] between nodes in the networks A and B (see Fig. 1.1a). The connections between interconnected nodes of the two networks are weighted by a factor p (see Fig. 1.1b),

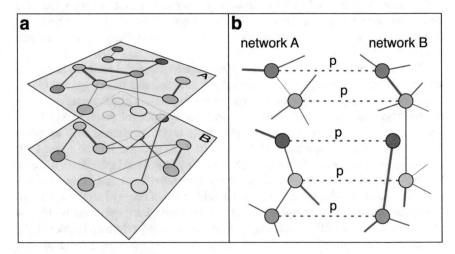

Fig. 1.1 (**a**) Schematic example of two interconnected networks A and B. In this representation, nodes of the same color are one-to-one interconnected. (**b**) In our model, inter-layer edges have weights equal to p (From Ref. [53])

any other weighted factor for the networks A and B is implicitly absorbed in their weights. The supra-adjacency matrix G of the whole network is therefore given by

$$G = \begin{pmatrix} A & p\mathbb{1} \\ p\mathbb{1} & B \end{pmatrix},$$ (1.1)

where $\mathbb{1}$ is the identity matrix with dimensions $N \times N$. Using this notation we can define the supra-laplacian of the interconnected network as

$$\mathcal{L} = \begin{pmatrix} \mathcal{L}_A + p\mathbb{1} & -p\mathbb{1} \\ -p\mathbb{1} & \mathcal{L}_B + p\mathbb{1} \end{pmatrix}.$$ (1.2)

The blocks present in $\mathcal{L}$ are square symmetric matrices of dimensions $N \times N$, In particular, $\mathcal{L}_A$ and $\mathcal{L}_B$ are the laplacians of the networks A and B, respectively.

Our investigation focuses on the analysis of the spectrum of the supra-Laplacian to ascertain the origin of the structural changes of the merging of networks in an interconnected system. The spectrum of the laplacian of a graph is a fundamental mathematical object for the study of the structural properties of the graph itself. There are many applications and results on graph Laplacian eigenpairs and their relations to numerous graph invariants (including connectivity, expanding properties, genus, diameter, mean distance, and chromatic number) as well as to partition problems (graph bisection, connectivity and separation, isoperimetric numbers, maximum cut, clustering, graph partition), and approximations for optimization problems on graphs (cutwidth, bandwidth, min-p-sum problems, ranking, scaling, quadratic assignment problem) [6, 13, 14, 43].

Note that, for any graph, all eigenvalues of its laplacian are non negative numbers. The smallest eigenvalue is always equal to zero and the eigenvector associated to it is trivially a vector whose entries are all identical. The second smallest eigenvalue λ_2 also called the *algebraic connectivity* [23] is one of the most significant eigenvalues of the Laplacian. It is strictly larger than zero only if the graph is connected. More importantly, the eigenvector associated to λ_2, which is called the *characteristic valuation* or *Fiedler vector* of a graph, provides even deeper information about its structure [24, 25, 45]. For example, the components of this vector associated to the various nodes of the network are used in spectral clustering algorithms for the bisection of graphs [49].

Our approach consists in the study of the behavior of the second smallest eigenvalue of the supra-laplacian matrix $\mathcal{L}$ and its characteristic valuation as a function of p, given the single-layer network laplacians $\mathcal{L}_A$ and $\mathcal{L}_B$. In the following, we will make use of the standard bra-ket notation for vectors. In this notation, $|x\rangle$ indicates a column vector, $\langle x|$ indicates the transposed (i.e., row vector) of $|x\rangle$, $\langle x|y\rangle = \langle y|x\rangle$ indicates the inner product between the vectors $|x\rangle$ and $|y\rangle$, $A|x\rangle$ indicates the action of matrix A on the column vector $|x\rangle$, and $\langle x|A$ indicates the action of matrix A on the row vector $\langle x|$. According to the theorem by Courant and Fisher (i.e., the so-called min-max principle) [16, 26], the second smallest eigenvalue of $\mathcal{L}$ is given by

$$\lambda_2 (\mathcal{L}) = \min_{|v\rangle \in \mathcal{V}} \langle v| \mathcal{L} |v\rangle , \tag{1.3}$$

where $|v\rangle \in \mathcal{V}$ is such that $\langle v|1\rangle = 0, \langle v|v\rangle = 1$. The vector $|1\rangle$ has $2N$ entries all equal to 1. Eq. (1.3) means that $\lambda_2 (\mathcal{L})$ is equal to the minimum of the function $\langle v| \mathcal{L} |v\rangle$, over all possible vectors $|v\rangle$ that are orthogonal to vector $|1\rangle$ and that have norm equal to one. The vector for which such minimum is reached is thus the characteristic valuation of the supra-laplacian (i.e., $\mathcal{L} |v\rangle = \lambda_2 |v\rangle$). We distinguish two blocks of size N in the vector $|v\rangle$ by writing it as $|v\rangle = |v_A, v_B\rangle$. In this notation, $|v_A\rangle$ is the part of the eigenvector whose components correspond to the nodes of network A, while $|v_B\rangle$ is the part of the eigenvector whose components correspond to the nodes of network B. We can now write

$$\langle v| \mathcal{L} |v\rangle = \langle v_A, v_B| \mathcal{L} |v_A, v_B\rangle = \\ \langle v_A| \mathcal{L}_A |v_A\rangle + \langle v_B| \mathcal{L}_B |v_B\rangle + \\ p \left(\langle v_A|v_A\rangle + \langle v_B|v_B\rangle - 2 \langle v_A|v_B\rangle \right)$$

and the previous set of constraints as $\langle v_A|1\rangle + \langle v_B|1\rangle = 0$ and $\langle v_A|v_A\rangle + \langle v_B|v_B\rangle = 1$, where now all vectors have dimension N. Accounting for such constraints, we can finally rewrite the minimization problem as

$$\lambda_2 (\mathcal{L}) = p + \min_{|v\rangle \in \mathcal{V}} \left\{ \langle v_A| \mathcal{L}_A |v_A\rangle + \langle v_B| \mathcal{L}_B |v_B\rangle - 2p \langle v_A|v_B\rangle \right\} . \tag{1.4}$$

First of all, we can simply state that the algebraic connectivity of Eq. (1.4) satisfies the inequality

$$\lambda_2 (\mathcal{L}) \le \frac{1}{2} \lambda_2 (\mathcal{L}_A + \mathcal{L}_B) , \tag{1.5}$$

where this upper bound comes out directly from the definition of the minimum of a function. For every $\mathcal{Q} \subseteq \mathcal{V}$, we have in fact that

$$\min_{|v\rangle \in \mathcal{V}} \langle v| \mathcal{L} |v\rangle \le \min_{|v\rangle \in \mathcal{Q}} \langle v| \mathcal{L} |v\rangle$$

simply because we are restricting the domain where looking for the minimum of the function $\langle v| \mathcal{L} |v\rangle$. The particular value of the upper bound of Eq. (1.5) is then given by setting $\mathcal{Q}$ as

$$|v\rangle = |v_A, v_B\rangle \in \mathcal{Q} \text{ is such that } |v_A\rangle = |v_B\rangle = |q\rangle , \\ \text{with } \langle q|1\rangle = 0, \langle q|q\rangle = 1/2.$$

Note that the upper bound of Eq. (1.5) does not depend on p, and thus represents the asymptotic value of $\lambda_2 (\mathcal{L})$ in the limit $p \to \infty$. This analytically proves the result established by Gómez et al. [30] through approximation methods.

The minimization problem of Eq. (1.4) can be solved using Lagrange multipliers. This means finding the minimum of the function

$$M = \langle v_A| \mathcal{L}_A |v_A\rangle + \langle v_B| \mathcal{L}_B |v_B\rangle - 2p \langle v_A|v_B\rangle$$
$$-r(\langle v_A|1\rangle + \langle v_B|1\rangle) - s(\langle v_A|v_A\rangle + \langle v_B|v_B\rangle - 1),$$

where the constraints of the minimization problem have been explicitly inserted in the function to minimize through the Lagrange multipliers r and s. In the following calculations, we will make use of the identities

$$\frac{\partial}{\partial |x\rangle} \langle t|x\rangle = \frac{\partial}{\partial |x\rangle} \langle x|t\rangle = \langle t|$$
$$\frac{\partial}{\partial |x\rangle} \langle x|x\rangle = 2\langle x|$$
$$\frac{\partial}{\partial |x\rangle} \langle x| A |x\rangle = 2\langle x| A, \text{ if } A = A^T$$

where $\frac{\partial}{\partial |x\rangle}$ indicates the derivative with respect to all the coordinates of the vector $|x\rangle$. Equating to zero the derivatives of M with respect to r and s, we obtain the constraints that we imposed. By equating to zero the derivative of M with respect to $|v_A\rangle$, we obtain instead

$$\frac{\partial M}{\partial |v_A\rangle} = 2\langle v_A| \mathcal{L}_A - 2p\langle v_B| - r\langle 1| - 2s\langle v_A| = \langle 0|, \tag{1.6}$$

and, similarly for the derivative of M with respect to $|v_B\rangle$, we obtain

$$\frac{\partial M}{\partial |v_B\rangle} = 2\langle v_B| \mathcal{L}_B - 2p\langle v_A| - r\langle 1| - 2s\langle v_B| = \langle 0|. \tag{1.7}$$

Multiplying both equations with $|1\rangle$, we have $2\langle v_A| \mathcal{L}_A |1\rangle - 2p\langle v_B|1\rangle - r\langle 1|1\rangle - 2s\langle v_A|1\rangle = 0$ and $2\langle v_B| \mathcal{L}_B |1\rangle - 2p\langle v_A|1\rangle - r\langle 1|1\rangle - 2s\langle v_B|1\rangle = 0$, that can be simplified in $2(p-s)\langle v_A|1\rangle - rN = 0$ and $2(p-s)\langle v_B|1\rangle - rN = 0$ because $\mathcal{L}_A|1\rangle = \mathcal{L}_B|1\rangle = |0\rangle$ and $\langle v_A|1\rangle = -\langle v_B|1\rangle$. Summing them, we obtain $r = 0$. Finally, we can write

$$(p-s)\langle v_A|1\rangle = 0$$
$$(p-s)\langle v_B|1\rangle = 0 . \tag{1.8}$$

These equations can be true in two cases: (i) $\langle v_A|1\rangle \neq 0$ or $\langle v_B|1\rangle \neq 0$ and $s = p$; (ii) $\langle v_A|1\rangle = \langle v_B|1\rangle = 0$. In the following, we analyze these two cases separately.

First, let us suppose that $s = p$, and that at least one of the two equations $\langle v_A|1\rangle \neq 0$ and $\langle v_B|1\rangle \neq 0$ is true. If we set $s = p$ in Eqs. (1.6) and (1.7), they become

$$\langle v_A| \mathcal{L}_A - p\langle v_B| - p\langle v_A| = \langle 0| \tag{1.9}$$

and

$$\langle v_B | \, \mathcal{L}_B - p \, \langle v_A | - p \, \langle v_B | = \langle 0 | \ . \qquad (1.10)$$

If we multiply the first equation with $|v_A\rangle$ and the second equation with $|v_B\rangle$, the sum of these two new equations is

$$\langle v_A | \, \mathcal{L}_A \, | v_A \rangle + \langle v_B | \, \mathcal{L}_B \, | v_B \rangle - 2p \, \langle v_A | v_B \rangle = p \ . \qquad (1.11)$$

If we finally insert this expression in Eq. (1.4), we find that the second smallest eigenvalue of the supra-laplacian is

$$\lambda_2 \, (\mathcal{L}) = 2p \ . \qquad (1.12)$$

We can further determine the components of Fiedler vector in this regime. If we take the difference between Eqs. (1.9) and (1.10), we have $\langle v_A | \, \mathcal{L}_A = \langle v_B | \, \mathcal{L}_B$. On the other hand, Eq. (1.12) is telling us that $\langle v_A | \, \mathcal{L}_A \, | v_A \rangle = - \langle v_B | \, \mathcal{L}_B \, | v_B \rangle$ because the only term surviving in Eq. (1.11) is the one that depends on p. Since $\langle v_A | \, \mathcal{L}_A \, | v_A \rangle$ ($\langle v_B | \, \mathcal{L}_B \, | v_B \rangle$) is always larger than zero, unless $|v_A\rangle = c\,|1\rangle$ ($|v_B\rangle = c\,|1\rangle$), with c arbitrary constant value, we obtain

$$|v_A\rangle = -\,|v_B\rangle \qquad \text{where } |v_A\rangle = \pm \frac{1}{\sqrt{2N}} \, |1\rangle \ . \qquad (1.13)$$

Thus in this regime, both the relations $\langle v_A | 1\rangle \neq 0$ and $\langle v_B | 1\rangle \neq 0$ must be simultaneously true. Eq. (1.13) also means that $\langle v_A | v_B\rangle = -\frac{1}{2}$.

The other possibility is that Eqs. (1.8) are satisfied because $\langle v_A | 1\rangle = 0$ and $\langle v_B | 1\rangle = 0$ are simultaneously true. In this case, the average value of the components of the vectors $|v_A\rangle$ and $|v_B\rangle$ is zero, i.e.,

$$\langle v_A | 1\rangle = \langle v_B | 1\rangle = 0 \ , \qquad (1.14)$$

and thus the coordinates of the Fiedler vector corresponding to the nodes of the same layer have alternatively negative and positive signs.

To summarize, the second smallest eigenvalue of the supra-laplacian matrix $\mathcal{L}$ is given by

$$\lambda_2 \, (\mathcal{L}) = \begin{cases} 2p & \text{, if } p \leq p^* \\ \leq \frac{1}{2} \lambda_2 \, (\mathcal{L}_A + \mathcal{L}_B) & \text{, if } p \geq p^* \end{cases} \ . \qquad (1.15)$$

Thus indicating that the algebraic connectivity of the interconnected system follows two distinct regimes, one in which its value is independent of the structure of the two layers, and the other in which its upper bound is limited by the algebraic connectivity of the weighted superposition of the two layers whose laplacian is given by $\frac{1}{2}(\mathcal{L}_A + \mathcal{L}_B)$. More importantly, the discontinuity in the first derivative of λ_2 is reflected in a radical change of the structural properties of the system happening at p^*, the tipping point. Such dramatic change is visible in the coordinates of characteristic valuation of the nodes of the two network layers. In the regime $p \leq p^*$, the components of the eigenvector are given by Eq. (1.13). This means that the two network layers are structurally disconnected and independent. For $p \geq p^*$, they instead obey Eq. (1.14), which means that the components of the vector corresponding to interconnected nodes of network A and B have the same sign, while nodes in the same layer have alternating signs. Thus in this second regime, the system connectivity is dominated by inter-layer connections, and the two network layers are structurally indistinguishable.

The tipping point p^* at which the transition occurs is the point at which we observe the crossing between the two different behaviors of λ_2, which means

$$p^* \leq \frac{1}{4} \lambda_2 (\mathcal{L}_A + \mathcal{L}_B) . \qquad (1.16)$$

Since inter-layer connections have weights that grows with p, the transition happens at the point at which the weight of the inter-layer connections exceeds the half part of the inverse of the algebraic connectivity of the weighted super-position of both network layers (see Fig. 1.2).

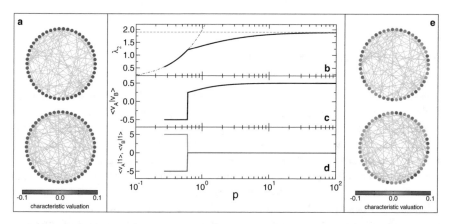

Fig. 1.2 Algebraic connectivity and Fiedler vector for two interconnected Erdős-Renyí networks of $N = 50$ nodes and average degree $\bar{k} = 5$. We consider a single realization of this model in which the critical point is $p^* = 0.602(1)$. (**a**) Characteristic valuation of the nodes in the two network layers for $p = 0.602$. (**b**) Algebraic connectivity of the system (*black line*). The discontinuity of the first derivative of λ_2 is very clear. The two different regimes $2p$ and $\frac{\lambda_2(\mathcal{L}_A+\mathcal{L}_B)}{2}$ are shown as *red dot-dashed* and *blue dashed lines*, respectively. (**c**) Inner product $\langle v_A | v_B \rangle$ between the part of the Fiedler eigenvector ($|v_A\rangle$) corresponding to nodes in the network A and the one ($|v_B\rangle$) corresponding to vertices in network B as a function of p. (**d**) Inner products $\langle v_A|1\rangle$ and $\langle v_B|1\rangle$ as functions of p. $\langle v_A|1\rangle$ and $\langle v_B|1\rangle$ indicate the sum of all components of the Fiedler vectors $|v_A\rangle$ and $|v_B\rangle$, respectively. (**e**) Characteristic valuation of the nodes in the two network layers for $p = 0.603$ (From Ref. [53])

Multiplex Networks Composed of More Than Two Layers

The results presented in the previous section can be extended, with analogous calculations, to multiplexes composed of ℓ network layers, with $\ell > 2$. The main result of Eq. (1.15) becomes

$$\lambda_2 \left(\mathcal{L} \right) = \begin{cases} \ell p & \text{, if } p \leq p^* \\ \leq \frac{1}{\ell} \lambda_2 \left(\sum_{i=1}^{\ell} \mathcal{L}_i \right) & \text{, if } p \geq p^* \end{cases}, \tag{1.17}$$

showing a discontinuity in the derivative of the algebraic connectivity at a certain value p^* estimated as

$$p^* \leq \frac{1}{\ell^2} \lambda_2 \left(\sum_{i=1}^{\ell} \mathcal{L}_i \right). \tag{1.18}$$

The algebraic connectivity of the multiplex can be thus written in terms of the algebraic connectivity of the superposition of all network layers that compose the multiplex. Unfortunately in the case of a multiplex network with $\ell > 2$, the Fiedler eigenvector cannot be fully characterized, and it is no longer possible to generalize Eqs. (1.13) and (1.14) to an arbitrary number of network layers.

Perturbed Multiplex Networks

The discontinuity in the first derivative of the algebraic connectivity λ_2 is due to the crossing between different eigenvalues in the spectrum of the supra-laplacian matrix $\mathcal{L}$. In the case of a multiplex composed of two layers, the presence of this crossing can be viewed as a simple consequence of the fact that the vector $|1, -1\rangle$ is always an eigenvector of the matrix in Eq. (1.2) for any value of p. By invoking the non crossing rule by von Neumann and Wigner [62], it is possible to show that the eigenvalue corresponding to this eigenvector, i.e., $\lambda = 2p$, must intersect all other eigenvalues of $\mathcal{L}$. Although in multiplex networks with an arbitrary number of coupled networks there is not a trivial eigenvector that can explain the crossing between eigenvalues, it is, however, worth asking if our results are still valid in presence of disorder. To this end, we consider a perturbed version of the matrix in Eq. (1.2). Essentially, instead of using the same p for every entry (i, j) of the off-diagonal blocks of Eq. (1.1), we use a weight of the type $p_{ij} = f(p, \epsilon_{ij})$ with ϵ_{ij} random variables such that the average $\langle f(p, \epsilon_{ij}) \rangle = p$. By applying this transformation, the vector $|1, -1\rangle$ is not longer an eigenvector $\mathcal{L}$ for every value of p. In presence of disorder, the non crossing rule by von Neumann and Wigner [62] tells us that no eigenvalues can cross as p varies. It is very interesting to note that the structural transition observed for the unperturbed case still holds, even for very small networks and not so small perturbations (see Fig. 1.3).

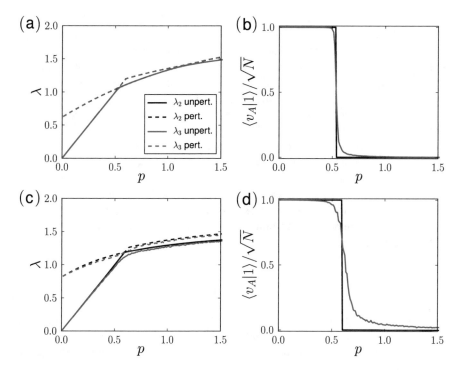

Fig. 1.3 Spectral properties of unperturbed and perturbed supra-laplacian matrices. The multiplex networks analyzed are composed of two Erdős-Rényi network models with $N = 50$ nodes and average degree $\bar{k} = 5$ as in the case of Fig. 1.2. (a) Second smallest λ_2 and third smallest λ_3 eigenvalues for the unperturbed network (*black lines*) as functions of the coupling strength p between different layers. We perturbed the coupling matrix by setting $p_{ij} = p + 10^{-1}(1/2 - \epsilon_{ij})$, with ϵ_{ij} uniform random variate in the interval $[0, 1]$. *Red lines* stand for average values of the second smallest and third smallest eigenvalues obtained over 100 perturbed realizations of the same starting network topology. (b) Sum of the eigenvector components corresponding to nodes in the same network layer, $\langle v_A | 1 \rangle$, for the unperturbed and the perturbed network of panel a. $\langle v_A | 1 \rangle$ is rescaled by the factor $\sqrt{N}$ to obtain values in the interval $[0, 1]$. (c) Same as in panel (a), but in this case the perturbation applied is $p_{ij} = p/2 + p\epsilon_{ij}$. (d) Same as in panel b, but for the perturbation scheme described in panel c

1.3 Conclusions

A physical interpretation of the algebraic structural transition that we are able to analytically predict can be given by viewing the function $\langle v | \mathcal{L} | v \rangle$ as an energy-like function. From this point of view, Eq. (1.3) becomes equivalent to a search for the ground state energy, and the characteristic valuation can be viewed as the ground state configuration. Such analogy is straightforward if one realizes that Eq. (1.3) is equivalent to the minimization of the weighted cut of the entire networked system [whose adjacency matrix G is defined in Eq. (1.1)], and that the minimum of this function corresponds to the ground state of a wide class of energy functions [40] and fitness landscapes [54]. These include, among others, the energy associated to the Ising spin models [44] and cost functions of combinatorial optimization problems, such as the traveling salesman problem [33]. In summary, the structural transition of interconnected networks involves a discontinuity in the first derivative of an energy-like function, and thus, according to the Ehrenfest classification of phase transitions, can be understood as a discontinuous transition [7].

Since the transition at the algebraic level has the same nature as the connectivity transition that has been studied by Buldyrev et al. in the same class of networked systems [10], it is worth to discuss about the relations between the two transitions. We can reduce our model to the annealed version of the model considered by Buldyrev et al. by setting $A = t^2A$, $B = t^2B$ and $p = t$, being $1 - t$ the probability that one node in one of the networks fails. All the results stated so far hold, with only two different interpretations. First, the upper bound of Eq. (1.16) becomes a lower bound for the critical threshold of the algebraic transition that reads in terms of occupation probability as

$$t_c \geq \frac{4}{\lambda_2 \left(\mathcal{L}_A + \mathcal{L}_B \right)} \, . \tag{1.19}$$

Second, the way to look at the transition must be reversed: network layers are structurally independent (i.e., the analogous of the non percolating phase) for values of $t \leq t_c$, while become algebraically connected (i.e., analogous of the percolating phase) when $t \geq t_c$.

As it is well known, the algebraic connectivity represents a lower bound for both the edge connectivity and node connectivity of graph (i.e., respectively the minimal number of edges or nodes that should be removed to disconnect the graph) [24]. Indeed, the algebraic connectivity of a graph is often used as a control parameter to make the graph more resilient to random failures of its nodes or edges [37]. Thus, the lower bound of Eq. (1.19) represents also a lower bound for the critical percolation threshold measured by Buldyrev et al. Interestingly, our prediction turns out to be a sharp estimate of the lower bound. For the Erdős-Rényi model, we have in fact $t_c \geq 2/\bar{k}$, if the two networks have the same average degree $\bar{k}$, and this value must be compared with $2.455/\bar{k}$ as predicted by Buldyrev et al. [10, 58]. Similarly, we are able to predict that t_c grows as the degree distribution of the network becomes more broad [14], in the same way as it has been numerically observed by Buldyrev et al. [10].

Although we are not able to directly map the algebraic transition to the percolation one, we believe that the existence of a first-order transition at the algebraic level represents an indirect support of the discontinuity of the percolation transition. We further emphasize that the transition is effectively present only if $t_c \leq 1$, and thus accordingly to Eq. (1.19) only if $\lambda_2 \left(\mathcal{L}_A + \mathcal{L}_B \right) \geq 4$. Such condition is verified for network layers that have a sufficiently large connectivity, and this qualitatively confirms the observation by Parshani et al. regarding a change in the nature of the percolation phase transition in interdependent networks with variable number of interdependent nodes [50].

In conclusion, we would like to briefly discuss the deep practical implications of our results. The abrupt nature of the structural transition is, in fact, not only visible in the limit of infinitely large systems, but for networks of any size, even if in presence of disorder. Thus, even real networked systems composed of few elements may be subjected to abrupt structural changes, including failures. Our theory provides, however, fundamental aids for the prevention of such collapses. It allows, in fact, not only the prediction of the critical point of the transition, but, more importantly, to accurately design the structure of such systems to make them more robust. For example, the percolation threshold of interconnected systems can be simply decreased by increasing the algebraic connectivity of the superposition of the network layers. This means that an effective strategy to make an interconnected system more robust is to avoid the repetition of edges among layers, and thus bring the superposition of the layers as close as possible to an all-to-all topology.

References

1. Albert, R., Barabási, A.-L.: Statistical mechanics of complex networks. Rev. Modern Phys. **74**(1), 47 (2002)
2. Albert, R., Jeong, H., Barabási, A.-L.: Error and attack tolerance of complex networks. Nature **406**(6794), 378–382 (2000)
3. Arenas, A., Diaz-Guilera, A., Kurths, J., Moreno, Y., Zhou, C.: Synchronization in complex networks. Phys. Rep. **469**(3), 93–153 (2008)
4. Barthélemy, M.: Spatial networks. Phys. Rep. **499**(1), 1–101 (2011)
5. Baxter, G.J., Dorogovtsev, S.N., Goltsev, A.V., Mendes, J.F.F.: Avalanche collapse of interdependent networks. Phys. Rev. Lett. **109**(24), 248701 (2012)
6. Bıyıkoglu, T., Leydold, J., Stadler, P.F.: Laplacian Eigenvectors of Graphs. Lecture Notes in Mathematics, vol. 1915. Springer, Berlin/Heidelberg (2007)
7. Blundell, S., Blundell, K.M.: Concepts in Thermal Physics. Oxford University Press, Oxford (2006)
8. Bollobás, B., Borgs, C., Chayes, J., Riordan, O.: Percolation on dense graph sequences. Ann. Probab. **38**(1), 150–183 (2010)
9. Brummitt, C.D., D'Souza, R.M., Leicht, E.A.: Suppressing cascades of load in interdependent networks. Proc. Natl. Acad. Sci. **109**(12), E680–E689 (2012)
10. Buldyrev, S.V., Parshani, R., Paul, G., Stanley, H.E., Havlin, S.: Catastrophic cascade of failures in interdependent networks. Nature **464**(7291), 1025–1028 (2010)
11. Castellano, C., Pastor-Satorras, R.: Thresholds for epidemic spreading in networks. Phys. Rev. Lett. **105**(21), 218701 (2010)
12. Cavers, M., Fallat, S., Kirkland, S.: On the normalized laplacian energy and general randić index r_{-1} of graphs. Linear Algebra Appl. **433**(1), 172–190 (2010)
13. Chung, F.R.K.: Spectral Graph Theory, vol. 92. American Mathematical Society, Providence (1997)
14. Chung, F.R.K., Lu, L., Vu, V.: Spectra of random graphs with given expected degrees. Proc. Natl. Acad. Sci. USA **100**(11), 6313–6318 (2003)
15. Cohen, R., Ben-Avraham, D., Havlin, S.: Percolation critical exponents in scale-free networks. Phys. Rev. E **66**(3), 036113 (2002)
16. Courant, R.: Über die eigenwerte bei den differentialgleichungen der mathematischen physik. Mathematische Zeitschrift **7**(1), 1–57 (1920)

17. De Domenico, M., Solè-Ribalta, A., Cozzo, E., Kivelä, M., Moreno, Y., Porter, M.A., Gòmez, S., Arenas, A.: Mathematical formulation of multi-layer networks. Phys. Rev. X **3**, 041022 (2013)
18. De Domenico, M., Solè-Ribalta, A., Gòmez, S., Arenas, A.: Navigability of interconnected networks under random failures. PNAS **111**(23), 8351–8356 (2014)
19. Dickison, M., Havlin, S., Stanley, H.E.: Epidemics on interconnected networks. Phys. Rev. E **85**(6), 066109 (2012)
20. Dorogovtsev, S.N., Goltsev, A.V., Mendes, J.F.: Critical phenomena in complex networks. Rev. Mod. Phys. **80**(4), 1275–1335 (2008)
21. Dorogovtsev, S.N., Mendes, J.F.F.: Evolution of networks. Adv. Phys. **51**(4), 1079–1187 (2002)
22. P.G. Doyle, Snell, J.L.: Random Walks and Electric Networks. Carus Mathematical Monographs, vol. 22. Mathematical Association of America, Washington, DC (1984)
23. Fiedler, M.: Algebraic connectivity of graphs. Czechoslov. Math. J. **23**(2), 298–305 (1973)
24. Fiedler, M.: A property of eigenvectors of nonnegative symmetric matrices and its application to graph theory. Czechoslov. Math. J. **25**(4), 619–633 (1975)
25. Fiedler, M.: Laplacian of graphs and algebraic connectivity. Banach Cent. Publ. **25**(1), 57–70 (1989)
26. Fischer, E.: Über quadratische formen mit reellen koeffizienten. Monatshefte für Mathematik **16**(1), 234–249 (1905)
27. Gao, J., Buldyrev, S.V., Havlin, S., Stanley, H.E.: Robustness of a network of networks. Phys. Rev. Lett. **107**(19), 195701 (2011)
28. Gao, J., Buldyrev, S.V., Stanley, H.E., Havlin, S.: Networks formed from interdependent networks. Nat. Phys. **8**(1), 40–48 (2012)
29. Goltsev, A.V., Dorogovtsev, S.N., Oliveira, J.G., Mendes, J.F.F.: Localization and spreading of diseases in complex networks. Phys. Rev. Lett. **109**(12), 128702 (2012)
30. Gomez, S., Diaz-Guilera, A., Gomez-Gardeñes, J., Perez-Vicente, C.J., Moreno, Y., Arenas, A.: Diffusion dynamics on multiplex networks. Phys. Rev. Lett. **110**(2), 028701 (2013)
31. Granell, C., Gómez, S., Arenas, A.: Dynamical interplay between awareness and epidemic spreading in multiplex networks. Phys. Rev. Lett. **111**, 128701 (2013)
32. Granell, C., Gómez, S., Arenas, A.: Competing spreading processes on multiplex networks: awareness and epidemics. Phys. Rev. E **90**, 012808 (2014)
33. Grover, L.K.: Local search and the local structure of NP-complete problems. Oper. Res. Lett. **12**(4), 235–243 (1992)
34. Harris, T.E.: The Theory of Branching Processes. Courier Dover Publications, New York (2002)
35. Hashimoto, K.: Zeta functions of finite graphs and representations of p-adic groups. In: Hashimoto, K., Namikawa, Y. (eds.) Automorphic Forms and Geometry of Arithmetic Varieties, pp. 211–280. Academic, Boston (1989)
36. Hu, Y., Ksherim, B., Cohen, R., Havlin, S.: Percolation in interdependent and interconnected networks: abrupt change from second-to first-order transitions. Phys. Rev. E **84**(6), 066116 (2011)
37. Jamakovic, A., Van Mieghem, P.: On the robustness of complex networks by using the algebraic connectivity. In: Das, A., Pung, H.K., Lee, F.B.S., Wong, L.W.C. (eds.) NETWORKING 2008 Ad Hoc and Sensor Networks, Wireless Networks, Next Generation Internet, 7th International IFIP-TC6 Networking Conference. Proceedings, Singapore, 5–9 May 2008. Lecture Notes in Computer Science, pp. 183–194. Springer (2008)
38. Kivelä, M., Arenas, A., Barthelemy, M., Gleeson, J.P., Moreno, Y., Porter, M.A.: Multilayer networks. J. Complex Netw. **2**(3), 203–271 (2014)
39. Klein, D.J., Randić, M.: Resistance distance. J. Math. Chem. **12**(1), 81–95 (1993)
40. Kolmogorov, V., Zabin, R.: What energy functions can be minimized via graph cuts? IEEE Trans. Pattern Anal. Mach. Intell. **26**(2), 147–159 (2004)
41. Krzakala, F., Moore, C., Mossel, E., Neeman, J., Sly, A., Zdeborová, L., Zhang, P.: Spectral redemption in clustering sparse networks. Proc. Natl. Acad. Sci. **110**(52), 20935–20940 (2013)

42. Liu, Y.-Y., Slotine, J.-J., Barabási, A.-L.: Controllability of complex networks. Nature **473**(7346), 167–173 (2011)
43. Merris, R.: Laplacian graph eigenvectors. Linear Algebra Appl. **278**(1), 221–236 (1998)
44. Mézard, M., Parisi, G., Virasoro, M.A.: Spin Glass Theory and Beyond, vol. 9. World Scientific, Singapore (1987)
45. Mohar, B., Alavi, Y.: The Laplacian spectrum of graphs. Graph Theory Comb. Appl. **2**, 871–898 (1991)
46. Newman, M.E.J.: The structure and function of complex networks. SIAM Rev. **45**(2), 167–256 (2003)
47. Newman, M.E.J.: Modularity and community structure in networks. Proc. Natl. Acad. Sci. **103**(23), 8577–8582 (2006)
48. Newman, M.E.J., Strogatz, S.H., Watts, D.J.: Random graphs with arbitrary degree distributions and their applications. Phys. Rev. E **64**(2), 026118 (2001)
49. Ng, A.Y., Jordan, M.I., Weiss, Y.: On spectral clustering analysis and an algorithm. In: Proceedings of Advances in Neural Information Processing Systems, vol. 14, pp. 849–856. MIT Press, Cambridge, MA (2001)
50. Parshani, R., Buldyrev, S.V., Havlin, S.: Interdependent networks: reducing the coupling strength leads to a change from a first to second order percolation transition. Phys. Rev. Lett. **105**(4), 048701 (2010)
51. Pastor-Satorras, R., Vespignani, A.: Epidemic spreading in scale-free networks. Phys. Rev. Lett. **86**(14), 3200 (2001)
52. Radicchi, F.: Driving interconnected networks to supercriticality. Phys. Rev. X **4**, 021014 (2014)
53. Radicchi, F., Arenas, A.: Abrupt transition in the structural formation of interconnected networks. Nat. Phys. **9**, 717–720 (2013)
54. Reidys, C.M., Stadler, P.F.: Combinatorial landscapes. SIAM Rev. **44**(1), 3–54 (2002)
55. Saumell-Mendiola, A., Serrano, M.Á., Boguñá, M.: Epidemic spreading on interconnected networks. Phys. Rev. E **86**(2), 026106 (2012)
56. Sole-Ribalta, A., De Domenico, M., Kouvaris, N.E., Diaz-Guilera, A., Gomez, S., Arenas, A.: Spectral properties of the Laplacian of multiplex networks. Phys. Rev. E **88**(3), 032807 (2013)
57. Son, S.-W., Bizhani, G., Christensen, C., Grassberger, P., Paczuski, M.: Percolation theory on interdependent networks based on epidemic spreading. EPL (Europhys. Lett.) **97**(1), 16006 (2012)
58. Son, S.-W., Grassberger, P., Paczuski, M.: Percolation transitions are not always sharpened by making networks interdependent. Phys. Rev. Lett. **107**(19), 195702 (2011)
59. Sood, V., Redner, S.: Voter model on heterogeneous graphs. Phys. Rev. Lett. **94**(17), 178701 (2005)
60. Szell, M., Lambiotte, R., Thurner, S.: Multirelational organization of large-scale social networks in an online world. Proc. Natl. Acad. Sci. **107**(31), 13636–13641 (2010)
61. Vazquez, F., Eguíluz, V.M.: Analytical solution of the voter model on uncorrelated networks. New J. Phys. **10**(6), 063011 (2008)
62. von Neumann, J., Wigner, E.: Non crossing rule. Zeitschrift für Physik **30**, 467–470 (1929)
63. Wilf, H.S.: generatingfunctionology. Academic (1990). Available at http://www.math.upenn.edu/~wilf/DownldGF.html

Chapter 2
Multilayer Networks: Metrics and Spectral Properties

Emanuele Cozzo, Guilherme Ferraz de Arruda, Francisco A. Rodrigues, and Yamir Moreno

Abstract Multilayer networks represent systems in which there are several topological levels each one representing one kind of interaction or interdependency between the systems' elements. These networks have attracted a lot of attention recently because their study allows considering different dynamical modes concurrently. Here, we revise the main concepts and tools developed up to date. Specifically, we focus on several metrics for multilayer network characterization as well as on the spectral properties of the system, which ultimately enable for the dynamical characterization of several critical phenomena. The theoretical framework is also applied for description of real-world multilayer systems.

2.1 Introduction

Complex network science relies on the hypothesis that the behavior of many complex systems can be explained by studying structural and functional relations among its components by means of a graph representation. The emergence of interconnected network models responds to the fact that complex systems include

E. Cozzo (✉)
Institute for Biocomputation and Physics of Complex Systems (BIFI), Zaragoza, Spain

Department of Theoretical Physics, University of Zaragoza, 50018 Zaragoza, Spain
e-mail: emcozzo@gmail.com

G.F. de Arruda • Francisco A. Rodrigues
Departamento de Matemática Aplicada e Estatística, Instituto de Ciências Matemáticas e de Computação, Universidade de São Paulo – Campus de São Carlos, Caixa Postal 668, 13560-970 São Carlos, SP, Brazil
e-mail: gui.f.arruda@gmail.com; francisco.rodrigues.usp@gmail.com

Y. Moreno
Institute for Biocomputation and Physics of Complex Systems (BIFI), Zaragoza, Spain

Department of Theoretical Physics, University of Zaragoza, 50018 Zaragoza, Spain

Complex Networks and Systems Lagrange Lab, Institute for Scientific Interchange, Turin, Italy
e-mail: yamir.moreno@gmail.com

© Springer International Publishing Switzerland 2016   17
A. Garas (ed.), *Interconnected Networks*, Understanding Complex Systems,
DOI 10.1007/978-3-319-23947-7_2

multiple subsystems organized as layers of connectivity. In this way, interconnected networks have emerged during the last few years as a general framework to deal with hyperconnected systems [1]. With the term interconnected networks one may refer to many types of connections among different networked systems: dependency relations among systems of different objects, cooperative or competitive relations among systems of different agents, or different channels of interactions among the same set of actors, to name a few. What these examples have in common is that different interaction modes among a differentiated or indistinguishable set of components/actors might exist.

Although this framework has been used for many years, only in the last several years it has attracted more attention and a number of formalisms have been proposed to deal with multilayer networks [2, 3]. Here we elaborate on a formalism developed recently and discussed at length in the review paper by Kivela et al. [4]. To this end, we report on a more refined formalism that is aimed at optimizing the study of a particular case of interconnected networks that is of much interest: Multiplex Networks.

In Multiplex Networks a set of agents might interact in different ways, i.e., through different means. Since a subset of agents is present at the same time in different networks of interactions (layers), these layers become interconnected. Examples of such type of systems can be founded in different fields, from biological systems, where the web of molecular interactions in a cell make use of many different biochemical channels and pathways, to technological systems, where person-to-person communication (usually machine-mediated) happens across many different modes. We take the last example as a paradigmatic one, which gave rise to the now popular term "hyperconnectivity" [5].

Suppose we are interested in analyzing a set of social agents (individuals, institutions, firms, etc.), who interact among them through a number of online social networks (OSNs) like Twitter, Facebook, etc. Some of these agents might be present in several OSNs and exchange information through them, using the information obtained in one network to communicate in another one, or integrating information across all of those in which they are active. We represent such a system as a set of graphs, one for each OSN, in which each actor who participates in it is represented by a node. These networks are the layers of the graph. In this scheme, the same actors are represented by a number of different nodes (as many nodes as the number of layers in which the actor is present). At the same time, we represent the fact that different nodes might denote the same actors, thus being related, by a coupling graph in which nodes representing the same actors are connected.

The rest of the chapter is organized as follows. The first section translates the aforementioned structural features in the formal language of graph theory. By doing that, we synthesize the topology of such a system in terms of matrices. In addition, as many years of research [6] have demonstrated, the relation between structure and function can be studied by means of the spectral properties of the matrices representing the graph structure. This is also studied in the second part of this chapter, where we give a simple example of the epidemic spreading process and analyze real world multilayer networks.

2.2 Notation, Basic Definitions and Properties

A multiplex network is a quadruple $\mathcal{M} = (\mathfrak{L}, \mathfrak{n}, \mathfrak{P}, \mathfrak{M})$. $\mathfrak{L} = \{1, \ldots, m\}$ is an index set that we call the layer set. Here we have assumed $\mathfrak{L} \subset \mathcal{N}$ for practical reasons and without loss of generality. We indicate the general element of $\mathfrak{L}$ with Greek lower case letters. Moreover, $\mathfrak{n}$ is a set of nodes and $\mathfrak{P} = (\mathfrak{n}, \mathfrak{L}, \mathfrak{N})$, $\mathfrak{N} \subseteq \mathfrak{n} \times \mathfrak{L}$ is a binary relation. Finally, the statement $(n, \alpha) \in \mathfrak{N}$ is read *node n participates in layer α*. We call the ordered pair $(n, \alpha) \in \mathfrak{N}$ a node-layer pair and we say that the node-layer pair (n, α) is the representative of node n in layer α.

On the other hand, $\mathfrak{M} = \{G_\alpha\}_{\alpha \in \mathfrak{L}}$ is a set of graphs, that we call layer-graphs, indexed by means of $\mathfrak{L}$. The node set of a layer-graph $G_\beta \in \mathfrak{M}$ is a sub-set $\mathfrak{n}_\beta \subset \mathfrak{N}$ such that $\mathfrak{n}_\beta = \{(n, \alpha) \in \mathfrak{P} \mid \alpha = \beta\}$, so the nodes of G_β are node-layer pairs; in that sense we say that node-layer pairs represent nodes in layers. The edge set of a graph $G_\alpha \in \mathfrak{M}$ is $\mathfrak{E}_\beta \subseteq \mathfrak{n}_\beta \times \mathfrak{n}_\beta$. Additionally, the binary relation $\mathfrak{P}$ can be identified with its graph $G_\mathfrak{P}$. $G_\mathfrak{P}$ has nodes set given by $\mathfrak{n} \cup \mathfrak{L}$, and edge set $\mathfrak{E}_\mathfrak{P} = \mathfrak{N}$, and we call it the *participation graph*.

Consider the graph $G_\mathfrak{C}$ on $\mathfrak{N}$ in which there is an edge between two node-layer pairs (n, α) and (m, β) only if $n = m$; that is, only if the two edges in the graph $G_\mathfrak{P}$ are incident on the same node $n \in \mathfrak{n}$, which means that the two node-layer pairs represent the same node in different layers. We call $G_\mathfrak{C}$ the coupling graph. It is easy to realize that the coupling graph is formed by $n = \mid \mathfrak{n} \mid$ disconnected components that are clicks or isolated nodes. Each clique is formed by all the representatives of a node in the layers, we call the components of $G_\mathfrak{C}$ *supra-nodes*.

Let's now also consider the graph $G_\mathfrak{l}$ on the same nodes set $\mathfrak{N}$, and in which there is an edge between two node-layer pairs (n, α), (m, β) only if $\alpha = \beta$; that is, only if the two edges in the graph $G_\mathfrak{P}$ are incident on the same node $\alpha \in \mathfrak{L}$. We call $G_\mathfrak{l}$ the layer graph. It is easy to realize that graph is formed by $m = \mid \mathfrak{L} \mid$ disconnected components that are clicks.

Finally, we can define the *supra-graph* $G_\mathcal{M}$ as the union of the layer-graphs with the coupling graph: $G_C \cup \mathfrak{M}$. $G_\mathcal{M}$ has node set $\mathfrak{N}$ and edge set $\bigcup_\alpha \mathfrak{E}_\alpha \cup \mathfrak{E}_\mathfrak{C}$. $G_\mathcal{M}$ is a synthetic representation of the Multiplex Network $\mathcal{M}$. It results that each layer-graph G_α is a sub-graph of $G_\mathcal{M}$ induced by $\mathfrak{n}_\alpha$. Furthermore, when all nodes participate in all layer-graphs the Multiplex Network is said to be fully aligned [4] and the coupling graph is made of n complete graphs of m nodes.

It is useful to come back to our system of social agents as a paradigmatic multiplex network to make sense of the previous definitions. The layer set is the list of OSNs, for example $\mathfrak{L} = \{Facebook, Twitter, Google+\}$. Since for practical purposes we want a set of indexes that are natural numbers, we may say that: Facebook is 1, Twitter is 2, and *Google+* is 3. The set of nodes is the set of social actors, for example $n = \{Marc, Alice, BiFi, Nick, Rose\}$. The binary relations represent the participation of each of these agents in some of the OSNs, thus we have that an statement of the type *Alice has a Facebook account* is represented by the pair *(Alice, 1)*, that is a node-layer pair. Each set of relation in each OSN is represented by a graph, for example the link *[(Alice, 1), (Nick, 1)]* means that Alice and Nick are friends on Facebook. If Alice has a Facebook account and a Twitter account, but not a Google+ account, in the coupling graph we will have the connected component $[(Alice, 1), (Alice, 2)]$ that is the supra-node related to Alice. If only the BiFi, Nick, and Rose have Google+ accounts, in the layer graph we will have the connected component $[(Bifi, 3), (Nick, 3), (Rose, 3)]$.

2.3 Multiplex Networks Related Matrices

Adjacency Matrices

In general, the adjacency matrix of a (unweighted, undirected) graph G with N nodes is a $N \times N$ (symmetric) matrix $\mathbf{A} = \{a_{ij}\}$, with $a_{ij} = 1$ only if there is an edge between i and j in G, and $a_{ij} = 0$ otherwise. We can consider the adjacency matrix of each of the graphs introduced in the previous section. The adjacency matrix of a layer graph G_α is a $n_\alpha \times n_\alpha$ symmetric matrix $\mathbf{A}^\alpha = a_{ij}^\alpha$, with $a_{ij}^\alpha = 1$ only if there is an edge between (i, α) and (j, α) in G^α. We call them layer adjacency matrices.

Likewise, the adjacency matrix of $G_\mathfrak{P}$ is an $n \times m$ matrix $\mathcal{P} = p_{i\alpha}$, with $p_{i\alpha} = 1$ only if there is an edge between the node i and the layer α in the participation graph, i.e. only if node i participate in layer α. We call it the participation matrix. The adjacency matrix of the coupling graph $G_\mathfrak{C}$ is an $N \times N$ matrix $\mathcal{C} = \{c_{ij}\}$, with $c_{ij} = 1$ only if there is an edge between node-layer pair i and j in $G_\mathfrak{C}$, i.e. if they are representatives of the same node in different layers. We can arrange the rows and the columns of $\mathcal{C}$ such that node-layer pairs of the same layer are contiguous and layers are ordered. We assume that $\mathcal{C}$ is always arranged in that way. It results that $\mathcal{C}$ is a block matrix with zero diagonal blocks. Thus, $c_{ij} = 1$, with $i, j = 1, \ldots, N$ represents an edge between a node-layer pair in layer 1 and a node-layer pair in layer 2 if $i < n_1$ and $n_1 < j < n_2$. Figure 2.1 shows a multiplex network and the respective matrices A and C.

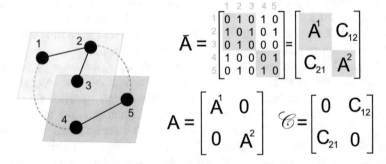

Fig. 2.1 Example of a multiplex network. The structure of each layer is represented by an adjacency matrix $\mathcal{A}^i$, where $i = 1, 2$. $\mathcal{C}_{lm}$ stores the connections between layers l and m. Note that the number of nodes in each layer is not the same

The Supra-adjacency Matrix

The *supra-adjacency* matrix is the adjacency matrix of the supra-graph $G_\mathcal{M}$. Just as $G_\mathcal{M}$, $\bar{A}$ is a synthetic representation of the whole multiplex $\mathcal{M}$. By definition, it can be obtained from the intra-layer adjacency matrices and the coupling matrix in the following way:

$$\bar{A} = \bigoplus_\alpha \mathbf{A}^\alpha + \mathcal{C}, \tag{2.1}$$

where the same consideration as in $\mathcal{C}$ applies for the indices. We also define $A = \bigoplus \mathbf{A}^\alpha$, and we call it the intra-layer adjacency matrix. Figure 2.1 shows the supra-adjacency and the intra-layer adjacency matrices of a multiplex network. Some basic metrics are easily calculated from the supra-adjacency matrix.

The degree of a node-layer i is the number of node-layers connected to it by an edge in $G_\mathcal{M}$ and is given by

$$K_i = \sum_j \bar{A}_{ij}. \tag{2.2}$$

Sometimes we write $i(\alpha)$ as an index, instead of simply i, to explicitly indicate that the node-layer i is in layer α even if the index i already uniquely indicates a node-layer pair. Since $\bar{A}$ can be read as a block matrix, with the $\mathbf{A}^\alpha$ on the diagonal blocks, the index $i(\alpha)$ can be interpreted as block index. It is also useful to define the following quantities

$$e_\alpha = \sum_{\beta < \alpha} n_\beta, \tag{2.3}$$

which we call the excess index of layer α. The layer degree of a node-layer i, $k_{i(\alpha)}$, is the number of neighbors it has in G^α, i.e., $k_{i(\alpha)} = \sum_j a_{ij}^\alpha$. By definition of $\bar{A}$

$$k_{i(\alpha)} = \sum_{j=1+e_\alpha}^{n_\alpha+e_\alpha} \bar{A}_{ij}. \tag{2.4}$$

The coupling degree of a node-layer i, $c_{i(\alpha)}$, is the number of neighbors it has in the coupling graph, i.e., $c_{i(\alpha)} = \sum_j c_{ij}$. From $\bar{A}$ we get

$$c_{i\alpha} = \sum_{\substack{j < e_\alpha, \\ j > n_\alpha + e_\alpha}} \bar{A}_{ij}. \tag{2.5}$$

Finally, we note that the degree of a node-layer can be expressed as

$$K_{i(\alpha)} = \sum_j \bar{A}_{ij} = k_{i\alpha} + c_{i\alpha}. \tag{2.6}$$

Equation (2.6) explicitly expresses the fact that the degree of a node-layer pair is the sum of its layer-degree plus its coupling-degree.

The Supra-Laplacian Matrix

Generally, the Laplacian matrix of a graph with adjacency matrix $\mathbf{A}$, or simply the Laplacian, is given by

$$\mathcal{L} = \mathcal{D} - \mathcal{A} \tag{2.7}$$

where $\mathbf{D} = diag(k_1, k_2, \dots)$ is the degree matrix.

Thus, it is natural to define the *supra-Laplacian* matrix of a Multiplex network as the Laplacian of its supra-graph

$$\bar{\mathcal{L}} = \bar{\mathcal{D}} - \bar{\mathcal{A}}, \tag{2.8}$$

where $\bar{\mathcal{D}} = diag(K_1, K_2, \dots, K_N)$ is the degree matrix. Besides, we can define the layer Laplacian of each graph G_α as

$$\mathbf{L}_\alpha = \mathbf{D}^\alpha - \mathbf{A}^\alpha, \tag{2.9}$$

and the Laplacian of the coupling graph

$$\mathcal{L}_C = \Delta - C \tag{2.10}$$

where $\Delta = diag(c_1, c_2, \dots, c_N)$ is the coupling-degree matrix.

By definition, we have

$$\bar{\mathcal{L}} = \bigoplus_\alpha \mathcal{L}^\alpha + \mathcal{L}_C. \tag{2.11}$$

Equation (2.11) takes a very simple form in the case of a node-aligned multiplex, i.e.,

$$\bar{\mathcal{L}} = \bigoplus_\alpha (\mathbf{L}^\alpha + cI_N) - \mathbf{K}_m \otimes I_n \tag{2.12}$$

where $\mathcal{K}_m$ is the adjacency matrix of a complete graph of m nodes, I_k is the $k \times k$ identity matrix and $c_i = c, \forall i \in \mathfrak{N}$ is the coupling degree of a node-layer pair.

Characteristic Matrices

2.3.1 Supra-nodes Characteristic Matrix

The supra-nodes characteristic matrix $\mathcal{S}_n = \{s_{ij}\}$ is an $N \times n$ matrix with $s_{ij} = 1$ only if the node-layer i is a representative of node j, i.e., it is in the connected component j in the graph $G_{\mathfrak{C}}$. We call it a characteristic matrix since supra-nodes partitions the node-layer set and S_n is the characteristic matrix of that partition.

2.3.2 Layers Characteristic Matrix

The layer characteristic matrix $\mathcal{S}_{\mathfrak{l}} = \{s_{ij}\}$ is an $N \times m$ matrix with $s_{ij} = 1$ only if the node-layer i is in the connected component j in the graph $G_{\mathfrak{l}}$. We call it a characteristic matrix since it is the characteristic matrix of the partition of the node-layer set induced by layers.

2.4 The Coarse-Grained Representation of a Multiplex Network

Nodes Partitions and Quotient graphs

We next briefly introduce the notion of network quotient associated to a partition of the node set. Suppose that $V_1, \ldots, V_m$ is a partition of the node set of a network G with adjacency matrix A, and write $n_i = |V_i|$. The quotient network Q of G is a coarse-grained representation of the network with respect to the partition. It has one node per cluster V_i and an edge from V_i to V_j weighted by an average connectivity from V_i to V_j

$$b_{ij} = \frac{1}{\sigma} \sum_{\substack{k \in V_i \\ l \in V_j}} a_{kl}. \tag{2.13}$$

Different choices are possible for the normalization parameter σ: $\sigma_i = n_i$, $\sigma_j = n_j$ or $\sigma_{ij} = \sqrt{n_i n_j}$. Depending on the choice for σ we call the resulting quotient respectively: left, right or symmetric quotient. We can express the left quotient $Q_l(A)$ in matrix form. Consider the $n \times m$ characteristic matrix of the partition $S = s_{ij}$, with $s_{ij} = 1$ if $i \in V_j$ and zero otherwise. Then

$$Q_l(A) = \Lambda^{-1} S^T A S, \tag{2.14}$$

where $\Lambda = diag\{n_1, \ldots, n_m\}$.

Aggregate Network and Network of Layers of a Multiplex Network

In the context of Multiplex Networks two quotient graphs arise naturally [7] by considering coupled node-layer pairs and layers. Supra-nodes partition the supra-graph, and the supra-nodes characteristic matrix S_n is the associated characteristic matrix. Then, we define the aggregate network of the multiplex network as the quotient associated to that partition:

$$\tilde{\mathbf{A}} = \Lambda^{-1} \mathcal{S}_n^T \bar{A} \mathcal{S}_n, \tag{2.15}$$

where $\Lambda = diag\{\kappa_1, \ldots, \kappa_n\}$ is the multiplexity degree matrix. Since, the Laplacian of the quotient is equal to the quotient of the Laplacian, the Laplacian of the aggregate network is given by:

$$\tilde{\mathbf{L}} = \Lambda^{-1} \mathcal{S}_n^T \bar{\mathcal{L}} \mathcal{S}_n. \tag{2.16}$$

In the same way, layers partition the supra-graph, thus the network of layers is defined by

$$\tilde{\mathbf{A}}_{\mathfrak{l}} = \Lambda^{-1} \mathcal{S}_{\mathfrak{l}}^T \bar{A} \mathcal{S}_{\mathfrak{l}}, \tag{2.17}$$

and its Laplacian is given by

$$\tilde{L}_{\mathfrak{l}} = \Lambda^{-1} S_{\mathfrak{l}}^T \bar{L} S_{\mathfrak{l}}. \tag{2.18}$$

2.5   Spectral Properties

The Largest Eigenvalue of $\bar{A}$

In the following we will interpret $\bar{A}$ as a perturbed version of $\mathcal{A}$, $\mathcal{C}$ being the perturbation. This choice is reasonable whenever

$$|| \mathcal{C} || < || \mathcal{A} ||. \tag{2.19}$$

Consider the largest eigenvalue λ of $\mathcal{A}$. Since $\mathcal{A}$ is a block diagonal matrix, the spectrum of $\mathcal{A}$, $\sigma(\mathcal{A})$, is

$$\sigma(\mathcal{A}) = \bigcup_{\alpha} \sigma(\mathbf{A}^\alpha), \tag{2.20}$$

$\sigma(\mathbf{A}^\alpha)$ being the spectrum of the adjacency-matrix $\mathbf{A}^\alpha$ of layer α. So, the largest eigenvalue λ of $\mathcal{A}$ is

$$\lambda = \max_\alpha \lambda_\alpha \tag{2.21}$$

with λ_α being the largest eigenvalue of $\mathbf{A}^\alpha$. We will look for the largest eigenvalue $\bar{\lambda}$ of $\bar{\mathcal{A}}$ as

$$\bar{\lambda} = \lambda + \Delta\lambda, \tag{2.22}$$

where $\Delta\lambda$ is the perturbation to λ due to the coupling C. For this reason, we call the layer δ for which $\lambda_\delta = \lambda$ the dominant layer. Let $\mathbf{1}_\alpha$ be a vector of size m with all entries equal to 0 except for the δ-th. If ϕ_δ is the eigenvector of $\mathbf{A}^\delta$ associated to λ_δ, we have that

$$\phi = \phi_\delta \otimes \mathbf{1}_\alpha \tag{2.23}$$

is the eigenvector associated to λ. Observe that ϕ have dimension n_δ, while $\mathbf{1}_\alpha$ have dimension m, where n_δ is the number of nodes on the dominant layer δ, yielding to a product of dimension $n_\delta \times m$, however it is not true if the number of nodes in is not the same on all layers. In such case we must construct the vector ϕ with zeros on all positions, except on the position of the leading eigenvector of the dominant layer. Then, we can approximate $\Delta\lambda$ as

$$\Delta\lambda \approx \frac{\phi^T C \phi}{\phi^T \phi} + \frac{1}{\lambda}\frac{\phi^T C^2 \phi}{\phi^T \phi}. \tag{2.24}$$

Because of the structure of ϕ and C, the first term on the r.h.s. is zero, while only the diagonal blocks of C^2 take part in the product $\phi^T C^2 \phi$. The diagonal blocks of C^2 are diagonals and

$$(C^2)_{ii} = \sum_{i'} C_{ii'} C_{i'i} = c_i. \tag{2.25}$$

Thus, we have that the perturbation is

$$\Delta\lambda \approx \frac{z}{\lambda}, \tag{2.26}$$

where we have defined the *effective multiplexity* z as the weighted mean of the coupling degree with the weight given by the squares of the entries of the leading eigenvector of A:

$$z = \sum_i c_i \frac{\phi_i^2}{\phi^T \phi}, \tag{2.27}$$

where $z = 0$ in a monoplex -single layer- network or $z = m - 1$ in a node aligned multiplex. Summing up, we have that the largest eigenvalue of the supra-adjacency matrix is equal to the largest eigenvalue of the dominant layer adjacency matrix at a first order approximation. As a consequence, for example, the critical point for an epidemic outbreak in a multiplex network is settled by that of the dominant layer at a first order approximation [8]. At second order, the deviation of $\bar{\lambda}$ from λ depends on the effective multiplexity and goes to zero with λ. See Figs. 2.2 and 2.3.

The approximation given in Eq. (2.26) can fail when the largest eigenvalue is near degenerated. We have two cases in which this can happen:

- the dominant layer is near degenerated,
- there is one (or more) layers with the largest eigenvalue near that of the dominant layer.

The accuracy of the approximation is related to the formula

$$\Delta\lambda \approx \phi^T C\phi + \sum_i \frac{(\phi^{(i)T} C\phi)}{\lambda - \lambda^{(i)}}, \qquad (2.28)$$

where $\lambda^{(i)}$ and $\phi^{(i)}$ are the non-dominant eigenvalues and the associated eigenvectors. In the first case it is evident that the second term on the r.h.s. will diverge, while in the latter, because of the structure of C, ϕ, and $\phi^{(i)}$, it is zero. In that case, we say that the multiplex network is near degenerated and we call the layers with the largest eigenvalues co-dominant layers.

When the multiplex network is near degenerated, ϕ used in the approximation of equation (2.26) has a different structure. Consider that we have l co-dominant layers δ_i, $i = 1, \ldots, l$. If ϕ_{δ_i} is the eigenvector of A^{δ_i} associated to λ_{δ_i}, we have that

$$\phi = \sum_{i=1}^{l} \phi_{\delta_i} \otimes \mathbf{1}_{\delta_i}. \qquad (2.29)$$

Note that the same comment on Eq. (2.23) also applies here. The term linear in C in the approximation of equation (2.26) is no more zero. We have

$$z_c = \frac{\phi^T C\phi}{\phi^T \phi} = \frac{1}{\phi^T \phi} \sum_{l,m:l\neq m} \phi_{\delta_l}^T \phi_{\delta_m} \qquad (2.30)$$

and we name z_c the correlated multiplexity. We can decompose z_c in the contribution of each single node-layer pair

$$z_{ci} = \frac{1}{\phi^T \phi} \sum_{m:m\neq l} \sum_j \phi_{\delta_l i} C_{ij} \phi_{\delta_m j}. \qquad (2.31)$$

Fig. 2.2 Effective multiplexity z as a function of the fraction of nodes coupled s for a two layers multiplex with 800 nodes with a power law distribution with $\gamma = 2.3$ in each layer. For each value of s, 40 different realizations of the coupling are shown while the intra-layer structure is fixed. In the panel on the top the z shows a two band structure, while in the panel on the bottom, it is continuous. The difference is due to the structure of the eigenvector

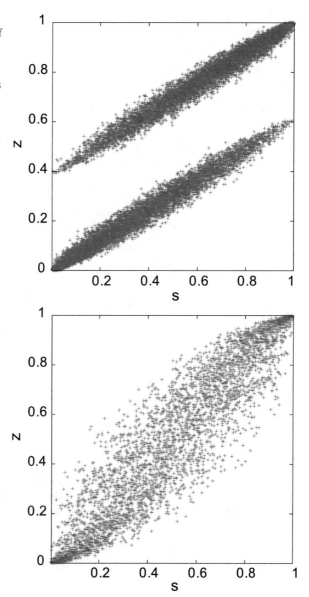

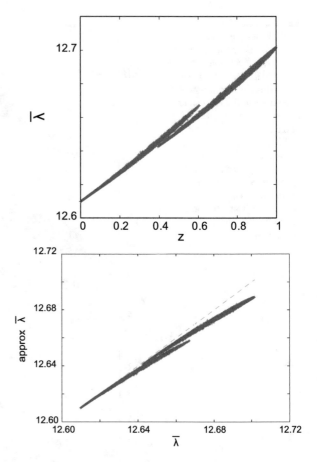

Fig. 2.3 Same setting of top panel of previous figure. On the *top*: calculated λ. We can see two branches corresponding to the two branches of the previous figure. *Bottom*: calculated vs approximated $\bar{\lambda}$

and we call z_{ci} the *correlated multiplexity degree* of node-layer i. By definition, coupled node-layer pairs have the same correlated multiplexity degree. So, if we have m_d co-dominant layers in the multiplex, we get

$$\Delta\lambda \approx z_c + \frac{z}{\lambda} = m_d \sum_{i\in\delta} z_{ci} + \frac{\sum_{i\in\delta} z_i}{\lambda}. \tag{2.32}$$

Spectral Relations Between Supra and Coarse-Grained Representations

The fundamental spectral result related to a quotient network is that adjacency eigenvalues of a quotient network interlace the adjacency eigenvalues of the parent network. That is, if $\mu_i, \ldots, \mu_m$ are the adjacency eigenvalues of the quotient network, and $\lambda_i, \ldots, \lambda_n$ are the adjacency eigenvalues of the parent network, it results that

$$\lambda_i \leq \mu_i \leq \lambda_{i+n-m}. \tag{2.33}$$

The same result applies for Laplacian eigenvalues. We can derive directly from that result a list of bounds for the supra-adjacency and the supra-Laplacian in terms of the aggregate network and of the network of layers [7]. Besides, in the case of node aligned multiplex networks, we have that the eigenvalues of the laplacian of the network of layers are a sub-set of the spectrum of the supra-Laplacian. This result is of special relevance in studying the structural properties of a multiplex network, since it states that the adjacency (Laplacian) eigenvalues of the coarse-grained representation of a multiplex interlace the adjacency (Laplacian) eigenvalues of the parent. In the case of a node-aligned multiplex, the Laplacian eigenvalues of the network of layers are a sub-set of the Laplacian eigenvalues of the parent Multiplex network.

The Second Eigenvalue of $\bar{L}$

A number of structural and dynamical properties of a network can be derived from the value of the first non-zero eigenvalue of the Laplacian. In the particular case of Multiplex Networks it has been shown that its behavior reflects a structural transition of the system [9]. We investigate the first non-zero eigenvalue of the supra-Laplacian of a node-aligned multiplex network. From the interlacing results of the previous section, we know that

$$\bar{\mu}_2 \leq \tilde{\mu}_{a_2} \tag{2.34}$$

and that

$$\bar{\mu}_2 \leq m. \tag{2.35}$$

m is always an eigenvalue of the supra-Laplacian, so, we can look for the condition under which $\bar{\mu}_2 = m$ holds. By combining equations (2.34) and (2.35), we arrive to the conclusion that if $m \geq \tilde{\mu}_{a_2}$, then $\bar{\mu}_2 \neq m$. On the other hand, we can approximate $\bar{\mu}_2$ as

$$\bar{\mu}_2 = \mu_2 + \Delta\mu_2, \tag{2.36}$$

where μ_2 is the eigenvalue of $\mathcal{L}$. We have

$$\Delta\mu_2 \approx \sum_{i<j} c_{ij}(x_i - x_j)^2, \tag{2.37}$$

where x_i are the elements of the eigenvector x associated to μ_2. Because of the structure of C and x, it results

$$\Delta\mu_2 \approx m - 1 \tag{2.38}$$

for a node aligned multiplex. Thus, since m is always an eigenvalue of $\bar{M}$, for that approximation to be correct, the following condition must hold

$$\mu_2 + m - 1 < m, \tag{2.39}$$

from which we can conclude that if $\mu_2 < 1$ then $\bar{\mu}_2 \neq m$.

In summary, we have that, if $\tilde{\mu}_{a2} < m$ or $\mu_2 < 1$ then $\bar{\mu}_2 \neq m$, but the converse is not true in general.

2.6 Applications

Dynamical Processes: Epidemic Spreading

An important application are the dynamical consequences of the interlacing properties on both adjacency and Laplacian matrices (see Sect. 2.5 and Ref. [7]). Here, as an example, we show the SIS epidemic spreading on the top of a multilayer network and the comparison with the aggregate network. Such dynamical process is based on the contact between individuals, or nodes, which can be infected or susceptible to the disease. Infected nodes, also called spreaders, spread the disease to its neighbors inside a time windows with probability β and recover from it with probability μ. Considering a discrete time approach, the Markov chain that formalizes this processes can be formally written by the iterative equation

$$p_i(t + 1) = \beta \sum_j \bar{A}_{ij} p_j(t) - \mu p_i(t), \tag{2.40}$$

where $p_i(t)$ is the probability of the node-layer pair i be infected at time t, $\bar{A}_{ij}$ are the elements of the supra-adjacency matrix $\bar{A}$, while β and μ are the infection and recovery probabilities, respectively. Such model consider the inter-layer and intra-layer as equal, which is a special case of the model presented in [8]. The critical point can be obtained by the first order approximation of Eq. (2.40) on its stationary regime, yielding

$$\beta_c = \frac{\mu}{\lambda_n(\bar{A})}, \tag{2.41}$$

where $\lambda_n(\bar{A})$ is largest eigenvalue of the supra-adjacency matrix $\bar{A}$ (see Eq. (2.1)). From the interlacing properties

$$\lambda_{n_\alpha}(\mathbf{A}^\alpha) \leq \lambda_n(\bar{A}), \tag{2.42}$$

Hence, the critical value β_c is bounded by the individual critical values and it is always lower or equal to the lowest individual layer critical value. In addition, observe that when the effective multiplexity, $z \approx 0$ in Eq. (2.27), the approximated leading eigenvalue of the multilayer supra-adjacency is given by the $\bar{\lambda} = \max\{\lambda(\bar{A})\}$. Furthermore, exploiting the network of layers spectra,

$$\lambda_m \leq \lambda_n(\bar{A}), \tag{2.43}$$

where λ_m is the largest eigenvalue of the network of layers, whose matrix is given by Eq. (2.17), implying another constraint to the critical point. In other words, the critical point of the network of layers bound from above the critical point of the multilayer.

Contrasting with the first model, now we consider a spreading process on the aggregate network, Eq. (2.15), hence

$$p_i(t+1) = \beta \sum_j \tilde{a}_{ij} p_j(t) - \mu p_i(t), \tag{2.44}$$

where $p_i(t)$ is the probability of the node i be infected at time t, $\tilde{a}_{ij}$ are the elements of the aggregated adjacency matrix $\tilde{\mathbf{A}}$, β is the infection probability and μ is the recovery probability. Observe that such process is different from the spreading described on Eq. (2.40), in which each node can infect its neighbors on any layer. On the other hand, in Eq. (2.44) each supra-node chooses a layer with uniform probability, than spreads the disease to all neighbors in that layer. Moreover, the critical point can be obtained using the same arguments as before, yielding to

$$\tilde{\beta}_c = \frac{\mu}{\lambda_n(\tilde{A})}, \tag{2.45}$$

where $\lambda_n(\tilde{\mathbf{A}})$ is largest eigenvalue of the aggregated adjacency matrix. Once again, for the interlacing results we have

$$\tilde{\beta}_c \geq \beta_c. \tag{2.46}$$

Such result imply that the spreading process on the multilayer structure is more efficient, or in the worst case as efficient as, than the process on the aggregate network [7].

The results of this section were formerly presented in [7]. In addition, it is noteworthy that a more complete model is proposed in [8], which consider the activity of the nodes and different spreading probabilities for the intra-layer and inter-layer edges. However, here we show the simplest cases, similar to the ones exposed in [8], in order to be more didactic. In spite of that, the examples shown here exemplify the importance of considering the multilayer structure and the role of the aggregated network and the network of layers.

Real-World Multilayer Networks

In order to evaluate real-world multilayer structures we study some networks available at *http://deim.urv.cat/~manlio.dedomenico/data.php*. We separate them into three different categories: (i) transportation networks; (ii) biological networks and (iii) social networks. We evaluate the maximum of the individual layer eigenvalues and the eigenvalue of the supra-adjacency matrix $\bar{A}$. Moreover, the approximations of the leading eigenvalues are also computed for comparison. Table 2.1 presents the results. Contrasting with monoplex systems, instead of one type of relationship, here we have m different types and also the connections between different layers. The average of the $k_{i\alpha}$ contains information about the relationship inside each layer, whereas the average of $c_{i\alpha}$ summarizes the relations between layers, i.e., between a given structure in two different contexts.

Regarding the networks studied here, we observe that biological networks tend to be sparser than social nets, specially considering the inter-layer relations. In addition, observe that there is a relationship between the average of the matrix C and the effective multiplexity z. For most of the networks, the first order approximation is accurate. However, some networks are better approximated by the second order approximation, for instance the CS. Furthermore, among all networks analyzed the only one that presented a poor approximation is the EU air transportation network, which can be explained by the high density of inter-layer couplings compared with the density of intra-layer connections.

2.7   Conclusion

The last years of research have just started to show that interconnected networks exhibit specific structural and dynamical properties that cannot be directly deduced from isolated networks. In order to gain understanding of such a system, a complete new toolbox is needed. On the other hand, such a new framework cannot be a naive extension of what has been developed for isolated, single layered, networks: we need that those tools be adapted to particular questions posed by interconnected networks. It is our conviction that the best way to tackle the problems ahead is to came back to the very basic concepts of graph theory and to build on them. The supra-adjacency matrix and the supra-Laplacian are examples of such basic objects, and the specific structural features of the interconnected system are reflected in them. In this way,

Table 2.1 Properties of real multilayer networks

| | Multilayer | Structure | | | | | | Approximation | | |
		N	m	$\langle k_{i\alpha} \rangle$	$\langle c_{i\alpha} \rangle$	$\langle K_i \rangle$	$\max\{\lambda(\bar{A})\}$	$\max_\alpha\{\lambda(A_\alpha)\}$	$\bar{\lambda} = \lambda + \frac{z}{\lambda}$	z	References
Transp.	London transport	369	3	2.211	0.155	2.366	3.787	3.782	3.786	0.004	[10]
	EU air transportation	450	37	3.528	11.417	14.945	30.274	19.315	19.690	7.226	[11]
Biological	C. elegans connectome	279	3	7.855	1.894	9.748	21.187	21.005	21.099	1.959	[12, 13]
	Danio Rerio genetic	155	5	1.900	0.367	2.267	4.941	4.391	4.484	0.408	[13, 14]
	Hepatitus C genetic	105	3	1.938	0.419	2.357	9.180	8.888	9.016	1.139	[13, 14]
	Homo genetic	18,222	7	8.570	1.516	10.086	119.311	119.286	119.308	2.541	[13, 14]
	Human Herpes4 genetic	216	4	1.847	0.444	2.291	12.484	12.329	12.453	1.530	[13, 14]
	Human HIV1 genetic	1005	5	2.115	0.393	2.508	16.427	16.227	16.364	2.226	[13, 14]
	Oryctolagus genetic	144	3	1.808	0.093	1.901	8.832	8.832	8.832	0.000	[13, 14]
	Xenopus genetic	461	5	1.935	0.488	2.423	6.203	6.093	6.189	0.583	[13, 14]
Social	CKM Physicians Innovation	246	3	4.065	1.884	5.950	8.125	6.703	6.998	1.975	[15]
	CS Aarhus	61	5	5.536	2.929	8.464	11.522	10.220	10.476	2.612	[16]
	Kapferer Tailor Shop	39	4	7.360	2.893	10.253	14.704	14.052	14.261	2.930	[17]
	Krackhardt High Tech	21	3	7.746	2.000	9.746	14.801	14.542	14.680	2.000	[18]
	Lazega Law Firm	71	3	15.670	1.991	17.660	24.194	23.868	23.952	1.994	[19]
	Vickers Chan 7th Graders	29	3	11.908	2.000	13.908	18.426	18.070	18.181	2.000	[20]

the rigorous study of these objects, as well as of their spectral properties, is likely to lead us to the correct understanding of the systems under study. Additionally we presented two applications, firstly the difference an epidemic spreading process that takes place on top of a multilayer or the aggregated network. Secondly, we have shown that perturbation theory is accurate enough when it comes to approximate the eigenvalue of a multilayer structure using the dominant (or co-dominant) layers.

Acknowledgements EC was supported by the FPI program of the Government of Aragon, Spain. This work has been partially supported by MINECO through Grant FIS2011-25167 to YM, Comunidad de Aragón (Spain) through FENOL to EC and YM; and by the EC FET-Proactive Project PLEXMATH (grant 317614) to YM. FAR and GAF thank Fapesp and CNPq for financial support give to this research.

References

1. Vespignani, A.: Complex networks: the fragility of interdependency. Nature **464**(7291), 984–985 (2010)
2. De Domenico, M., Solé-Ribalta, A., Cozzo, E., Kivela, M., Moreno, Y., Porter, M.A., Arenas, A.: Mathematical formulation of multilayer networks. Phys. Rev. X **3**(4), 041022 (2013)
3. Boccaletti, S., Bianconi, G., Criado, R., del Genio, C.I., Gómez-Gardeñes, J., Romance, M., Sendiña-Nadal, I., Wang, Z., Zanin, M.: The structure and dynamics of multilayer networks. Phys. Rep. **544**(1), 1–122 (2014)
4. Kivela, M., Arenas, A., Barthelemy, M., Gleeson, J.P., Moreno, Y., Porter, M.A.: Multilayer networks. J. Complex Netw. **2**(3), 203–271 (2014)
5. Wellmann, B.: Physical place and cyberplace: the rise of networked individualism. Int. J. Urban Reg. Res. **1**, 227–252 (2001)
6. Boccaletti, S., Latora, V., Moreno, Y., Chavez, M., Hwang, D.U.: Complex networks: structure and dynamics. Phys. Rep. **424**(4), 175–308
7. Sánchez-García, R.J., Cozzo, E., Moreno, Y.: Dimensionality reduction and spectral properties of multilayer networks. Phys. Rev. E **89**(5), 052815
8. Cozzo, E., Banos, R.A., Meloni, S., Moreno, Y.: Contact-based social contagion in multiplex networks. Phys. Rev. E **88**(5), 050801
9. Radicchi, F., Arenas, A.: Abrupt transition in the structural formation of interconnected networks. Nat. Phys. **9**, 717–720 (2013)
10. De Domenico, M., Sole-Ribalta, A., Gomez, S., Arenas, A.: Navigability of interconnected networks under random failures. PNAS **111**, 8351–8356 (2014)
11. Cardillo, A.: Gomez-Gardenes, J., Zanin, M., Romance, M., Papo, D., del Pozo, F., Boccaletti, S.: Emergence of network features from multiplexity. Sci. Rep. **3** (2013). http://www.nature.com/articles/srep01344
12. Chen, B.L., Hall, D.H., Chklovskii, D.B.: Wiring optimization can relate neuronal structure and function. PNAS **103**(12), 4723–4728 (2006)
13. De Domenico, M., Porter, M.A., Arenas, A.: MuxViz: a tool for multilayer analysis and visualization of networks. J. Complex Netw. (2014). doi:10.1093/comnet/cnu038
14. Stark, C., Breitkreutz, B.-J., Reguly, T., Boucher, L., Breitkreutz, A., Tyers, M.: Biogrid: a general repository for interaction datasets. Nucleic Acids Res. **34**(1), D535–D539 (2006)
15. Coleman, J., Katz, E., Menzel, H.: The diffusion of an innovation among physicians. Sociometry **20**, 253–270 (1957)
16. Magnani, M., Micenkova, B., Rossi, L.: Combinatorial analysis of multiple networks. arXiv:1303.4986

17. Kapferer, B.: Strategy and Transaction in an African Factory: African Workers and Indian Management in a Zambian Town. Manchester University Press, Manchester (1972)
18. Krackhardt, D.: Cognitive social structures. Soc. Netw. **9**, 104–134 (1987)
19. Snijders, T.A.B., Pattison, P.E., Robins, G.L., Handcock, M.S.: New specifications for exponential random graph models. Sociol. Methodol. **36**, 99–153 (2006)
20. Vickers, M., Chan, S.: Representing Classroom Social Structure. Victoria Institute of Secondary Education, Melbourne (1981)

Chapter 3
An Ensemble Perspective on Multi-layer Networks

Nicolas Wider, Antonios Garas, Ingo Scholtes, and Frank Schweitzer

Abstract We study properties of multi-layered, interconnected networks from an ensemble perspective, i.e. we analyze ensembles of multi-layer networks that share similar aggregate characteristics. Using a diffusive process that evolves on a multi-layer network, we analyze how the speed of diffusion depends on the aggregate characteristics of both intra- and inter-layer connectivity. Through a block-matrix model representing the distinct layers, we construct transition matrices of random walkers on multi-layer networks, and estimate expected properties of multi-layer networks using a mean-field approach. In addition, we quantify and explore conditions on the link topology that allow to estimate the ensemble average by only considering aggregate statistics of the layers. Our approach can be used when only partial information is available, like it is usually the case for real-world multi-layer complex systems.

3.1 Introduction

Networks are often used to describe interactions among the elements of a complex system. But until recently, the standard assumption in the literature was that networks are isolated entities and do not interact with other networks. Today we understand that this assumption is a rough simplification, since real networks usually have complex patterns of interaction with other networks. In order to study more realistic systems, network theory extended its perspective to account for these network to network interactions, and to investigate their influence on various processes of interest that may use the network topology as substrate [1–4].

Networks consisting of multiple networks and the connections between them are called *interconnected* or *multi-layer* networks [5]. In a multi-layered network each individual layer contains a network that is different from the networks contained in other layers, and the layer interconnectivity refers to the fact that nodes in different layers can be connected to each other. Nevertheless, it is often possible to extend

N. Wider (✉) • A. Garas • I. Scholtes • F. Schweitzer
Chair of Systems Design, ETH Zürich, Weinbergstrasse 58, 8092, Zürich, Switzerland
e-mail: nwider@ethz.ch; agaras@ethz.ch; ischoltes@ethz.ch; fschweitzer@ethz.ch

© Springer International Publishing Switzerland 2016
A. Garas (ed.), *Interconnected Networks*, Understanding Complex Systems,
DOI 10.1007/978-3-319-23947-7_3

and apply methods developed for single-layer (isolated) networks to multi-layer networks, assuming that all layers and the connections between them are known precisely. Unfortunately, when creating networks using relational data on real-world systems we are often confronted with situations where we *lack information* about the details of their multi-layer structure. In such situations, ensemble-based approaches allow us to reason about the expected properties of such networks, provided that we have access to aggregate statistics which can be used to define a statistical ensemble.

For instance, there are situations in which we are able to precisely map the topology *within* each layer individually, but we may not be able to obtain the detailed topology of connections *across* different layers. As an example, we may consider the topology of connections between users in different online social networks (OSNs). Such a system can be represented as a multi-layer network, where each layer represents the network of connections between users *within* one OSN. In addition, cross-layer connections are due to users which are members of multiple OSNs at the same time, and which can thus drive the dissemination of information across OSNs. Data on the network topology within particular OSNs are often readily available, however it is in general very difficult to identify accounts of the same user in different OSNs.

Contrary to the situation described above, we may also consider situations in which detailed information on the topology of cross layer links is available, while the detailed topology of connections *within* layers is not known. For example, there may be a rather small number of static links *across* layers, while the topology of links *within* layers is too large and too dynamic to allow for a detailed mapping. Again, in such a situation we may still have access to partial, aggregate information on the *inter-layer* connectivity (such as the number of nodes or the density of links) which we can use in order to reason about a multi-layer complex system.

Both the above situations lead to multi-layer networks, and both require us to reason about a system with incomplete information. This problem can be addressed from a macroscopic perspective using *statistical ensembles*, and in this chapter we extend the ensemble perspective to multi-layer networks, where we have access to mere aggregate statistics either on links *within* or *across* layers. Combining both detailed and aggregate information on the links in a multi-layer network, we first define a statistical ensemble, i.e. a probability space containing all network realizations that are consistent with available information. Secondly, we assume a probability mass function which assigns a probability to each possible realization in the ensemble. And finally, using either analytical or numerical techniques, we use the resulting probability space to reason about the *expected properties* of a network given that it is drawn from the ensemble. The rest of the chapter is structured as follows. In Sect. 3.2 we present our methodological approach to model ensembles of multi-layer networks, we formally introduce the diffusion process that is assumed to run on the multi-layer network, and we introduce a method that allows to aggregate the statistics of links inside layers and across layers. In Sect. 3.3 we introduce a mean-field approach to approximate ensemble averages, and we investigate under which conditions it can be used to argue about diffusion in multi-layer networks. In particular we discuss three distinct cases according to different levels of information that we may have about the topology of links across the layers or inside the layers.

3.2 Methods and Definitions

In our analysis we investigate a diffusion process that evolves on a static multi-layer network. More precisely we focus on diffusion dynamics modeled by a random walk process. In the following we provide a short summary that highlights the important properties of this process. Note that these are facts already known from the theory of random walks on single layer networks [6, 9], but later on we will extend this framework to multi-layer networks.

We assume a discrete time random walk process on a network $\mathbf{G}$ that consist of n nodes. Starting at an arbitrary node, at each step of the process the walker moves to an adjacent node, so that for a pair of nodes i, j the probability for a walker to move from node i to node j is given by the corresponding entry $\mathbf{T}_{ij}$ of a *transition matrix* $\mathbf{T}$. Since we have $\sum_j P(i \rightarrow j) = 1$, the transition matrix is row stochastic.

We further consider a vector $\pi^t \in \mathbf{R}^n$, whose entries π_i^t indicate the probability of a random walker to visit node i after t steps of the process. Here, we consider π^0 as a given initial distribution, whose entries π_i^0 give the probability that the random walker has started at node i. The change of visitation probabilities $\pi^t \rightarrow \pi^{t+1}$ can then be calculated based on the transition matrix as follows:

$$\pi^{t+1} = \pi^t \mathbf{T}. \tag{3.1}$$

Since this is an iterative process starting with π_0, the visitation probability vector after t time steps can be calculated as $\pi^t = \pi^0 \mathbf{T}^t$, and we can investigate the long-term behavior of the random walk process for $t \rightarrow \infty$. For a visitation probability vector π^* such that $\pi^* \mathbf{T} = \pi^*$, we can say that the process reaches a stationary distribution π^*, and if the transition matrix $\mathbf{T}$ is primitive, the Perron-Frobenius theorem guarantees that such a unique stationary distribution π^* exist.

In order to assess the convergence time of a random walk process, we can study the *total variation distance* between visitation probabilities π^t after t steps and the stationary distribution π^*. For two distributions π and π', the total variation distance is defined according to Ref. [7] as

$$\Delta(\pi, \pi') := \frac{1}{2} \sum_i |\pi_i - \pi_i'|, \tag{3.2}$$

where π_i indicates the i-th entry of π.

As a proxy for diffusion speed, we can now investigate how long it takes until the total variation distance $\Delta(\pi^t, \pi^*)$ falls below some given threshold value ϵ (for a small ϵ). In other words, we study how many steps $t(\epsilon)$ a random walker needs such that $\Delta(\pi^t, \pi^*) \leq \epsilon$ for $t \geq t(\epsilon)$

The eigenvalues $1 = \lambda_1 \geq |\lambda_2| \geq \ldots \geq |\lambda_n|$ of a row-stochastic matrix necessarily have absolute values that fall between zero and one, while the largest eigenvalue λ_1 is necessarily one. The number of required time steps $t(\epsilon)$ (and thus the diffusion speed of the random walk process) can be estimated by means of the second-largest eigenvalue λ_2 of $\mathbf{T}$,

$$t(\epsilon) \sim \frac{-1}{\ln(|\lambda_2|)}. \tag{3.3}$$

For a detailed derivation see Ref. [8]. Equation (3.3) shows that a second-largest eigenvalue λ_2 close to one implies slow convergence, while λ_2 close to zero implies fast convergence. Therefore in the following we use the second-largest eigenvalue of a transition matrix $\lambda_2(\mathbf{T})$ as a proxy to measure and quantify the convergence behavior on a network.

Multi-layer Network

The purpose of our study is to investigate diffusion processes on ensembles of networks with multiple interconnected layers. Thus, in the following we briefly introduce the notion of multi-layer networks used in this chapter. Let us consider a multi-layer network denoted by $\mathbf{G}$ that consist of L non-overlapping layers $G_1, \ldots, G_L$. Each of these layers G_l is a single-layer network $G_l = (V_l, E_l)$ where $V(G_l)$ and $E(G_l)$ denote the nodes and links of layer l respectively. We call the links $E(G_l)$ between nodes *within* the layers l *intra*-links. The multi-layer network $\mathbf{G}$ consist in total of n nodes, where $n = \sum_{l=1}^{L} |V(G_l)|$. In addition, we assume a set $E_I(\mathbf{G})$ of *inter-layer* links which connect nodes *across* layers, i.e. for each edge $(u, v) \in E_I$ we have $u \in V(G_i)$ and $V(G_j)$ for $i \neq j$. Inter-layer links induce a multipartite network with the independent sets $G_1, \ldots, G_L$.

In our study we consider undirected and unweighted networks, however some of our results may hold even for directed or weighted networks. Furthermore, from the perspective of a random walk process, we assume that inter- and intra-layer links are indistinguishable, i.e. transitions are made purely randomly irrespective of the type of link. As such, the multi-layer network can also be viewed as a huge single network consisting of subgraphs $G_1, \ldots, G_L$.

As mentioned above diffusion dynamics on networks can be studied analytically using transition matrices of random walkers [9, 10]. The multi-layer structure of a network can explicitly be incorporated in a random walk model by constructing a so-called *supra*-transition matrix [3, 11] similar to the supra-adjacency matrix used in [10, 12, 13]. The supra-adjacency matrix of a multi-layer network $\mathbf{G}$ can be defined in a block-matrix structure as

$$
\mathbf{A} = \begin{pmatrix}
\mathbf{A}_1 & \cdots & \mathbf{A}_{1t} & \cdots & \mathbf{A}_{sL} \\
\vdots & \ddots & \vdots & \ddots & \vdots \\
\mathbf{A}_{s1} & \cdots & \mathbf{A}_{st} & \cdots & \mathbf{A}_{sL} \\
\vdots & \ddots & \vdots & \ddots & \vdots \\
\mathbf{A}_{L1} & \cdots & \mathbf{A}_{Lt} & \cdots & \mathbf{A}_L
\end{pmatrix} . \tag{3.4}
$$

On the diagonal we have the adjacency matrices $\mathbf{A}_1, \ldots, \mathbf{A}_L$ corresponding to the layers $G_1, \ldots, G_L$, thus entries of these block matrices represent the *intra-layer links* of the multi-layer network. Off-diagonal matrices $\mathbf{A}_{ij}$ for $i, j \in \{1, \ldots, L\}$ with $i \neq j$ represent *inter-layer* links that connect nodes in layer G_i to nodes in layer G_j. Since we consider undirected networks we have $\mathbf{A}_{ij}^\top = \mathbf{A}_{ji}$.

Based on a supra-adjacency matrix $\mathbf{A}$ we can easily define a *supra-transition* matrix $\mathbf{T}$ of a random walker on a multi-layer network G. In block-matrix form such a matrix can be written as:

$$
\mathbf{T} = \begin{pmatrix}
\mathbf{T}_1 & \cdots & \mathbf{T}_{1t} & \cdots & \mathbf{T}_{sL} \\
\vdots & \ddots & \vdots & \ddots & \vdots \\
\mathbf{T}_{s1} & \cdots & \mathbf{T}_{st} & \cdots & \mathbf{T}_{sL} \\
\vdots & \ddots & \vdots & \ddots & \vdots \\
\mathbf{T}_{L1} & \cdots & \mathbf{T}_{Lt} & \cdots & \mathbf{T}_L
\end{pmatrix} . \tag{3.5}
$$

Here, each entry T_{ij} is defined as:

$$
T_{ij} = \frac{a_{ij}}{\sum_{k=1}^{n} a_{kj}} , \tag{3.6}
$$

where a_{ij} are the corresponding entries of the supra-adjacency matrix $\mathbf{A}$. Note that, due the presence of inter-layer links, block matrices $\mathbf{T}_{ij}$ are in general not equal to the row-normalized version of block matrices $\mathbf{A}_{ij}$. The supra transition matrix defined above can be used to model a random walk process on a multi-layer network.

From an analytical perspective the supra-transition matrix can be treated in the same way as the transition matrix of a single layer as explained above. In the case of undirected networks the eigenvalues of a transition matrix are related to the eigenvalues of the normalized Laplacian matrix. In our case we study the second-largest eigenvalue of the supra-transition matrix and use it as a proxy for the efficiency of a network with respect to a diffusion process as pointed out above.

Using $\mathbf{T}$ we are able to model a diffusion process on a multi-layer network. Since we especially want to emphasize the relevance of the inter-links, in the next section we introduce a transition matrix that only considers transitions across layers and not between individual nodes. As we will see later, this aggregated transition matrix is useful to distinguish the influence of inter-layer and intra-layer links on the convergence behavior of a random walk process.

Multi-layer Aggregation

The supra-transition matrix **T** introduced previously contains transition probabilities for any pair of nodes in the multi-layer network. In this sense **T** could also be the transition matrix of a large network, which is not divided in separate layers. In order to understand the effects of a layered structure, in this section we focus explicitly on transitions across layers. To do this we aggregate the statistics of inter-links and the intra-links of all single layers, and thus, we homogenize all individual nodes that belong to the same layer. This way we reduce the supra-transition matrix **T** of dimension n to an aggregated transition matrix $\mathfrak{T}$ of dimension L. We call this process *multi-layer aggregation* and the matrix $\mathfrak{T}$ the *layer-aggregated* or just *aggregated* transition matrix. Later on we will provide a relation between the eigenvalues of **T** and $\mathfrak{T}$, which will allow us to decompose the spectrum of **T**. This is important since the convergence behavior of a random walk process depends on the second largest eigenvalues of **T**.

Let us begin by discussing the construction process of the layer-aggregated transition matrix. Our goal is to define transition probabilities across any two layers G_s and G_t by averaging the transitions between any two nodes of G_s and G_t. Under certain conditions which will be specified in the following, these average transition probabilities can be representative for all nodes of the different layers.

Let **G** be a multi-layer network that consists of L layers $G_1, \ldots, G_L$. The transition probability to go from node v_i to any node v_j in **G** is defined as

$$P(v_i \rightarrow v_j) = \frac{\omega(v_i, v_j)}{\sum_k \omega(v_i, v_k)} \tag{3.7}$$

where $\omega(v_i, v_j)$ is the weight of a link connecting v_i with v_j. This is a general formalism, but since we only consider unweighted networks we have $\omega(v_i, v_j) = 1$ if and only if there is a link between the nodes v_i and v_j.

For each node v_i in layer G_s we require that the transition probabilities $P(v_i \rightarrow *)$ to nodes in another layer G_t fulfill the following equation

$$\alpha_{ss} \sum_{v_j \in V(G_s)} P(v_i \rightarrow v_j) = \alpha_{st} \sum_{v_k \in V(G_t)} P(v_i \rightarrow v_k) \quad \forall v_i \in V(G_s), \tag{3.8}$$

where α_{st} is a factor that only depends on the layers G_s and G_t. The factor α_{ss} is used to normalize the transitions, such that $\sum_t \alpha_{st} = 1$ is satisfied. In other words Eq. (3.8) implies that the probability for a random walker at node i to stay inside layer G_s is a multiple of the probability to switch to layer G_t.

We can see that α_{st} is independent of i, and therefore $\mathbf{T}_{st} = \alpha_{st} \mathbf{R}_{st}$ where $\mathbf{R}_{st}$ is a row stochastic matrix. This means that $\mathbf{T}_{st}$ resembles a scaled transition matrix, and α_{st} represents the weighted fraction of all links starting in G_s that end up in G_t. Thus, we can define the aggregation of a supra-adjacency matrix satisfying Eq. (3.8) as

$$
\mathfrak{T} = \left(
\begin{array}{ccc|ccc|c}
\alpha_{11} & \cdots & \alpha_{1t} & \cdots & \alpha_{sL} \\
\vdots & \ddots & \vdots & \ddots & \vdots \\
\hline
\alpha_{s1} & \cdots & \alpha_{st} & \cdots & \alpha_{sL} \\
\vdots & \ddots & \vdots & \ddots & \vdots \\
\hline
\alpha_{L1} & \cdots & \alpha_{Lt} & \cdots & \alpha_{LL}
\end{array}
\right) . \tag{3.9}
$$

If a multi-layer network $\mathbf{G}$ satisfies Eq. (3.8) we can follow that the spectrum of the aggregated matrix $\mathfrak{T} = \{\alpha_{st}\}_{st}$ is

$$
Spec(\mathfrak{T}) = \{1, \lambda_2, \ldots, \lambda_L\}, \tag{3.10}
$$

and it holds that $\lambda_2, \ldots, \lambda_L \in Spec(\mathbf{T})$ (see Proposition 1 in the Appendix).

This relation implies that the aggregated matrix $\mathfrak{T}$ preserves L eigenvalues of the supra-transition matrix $\mathbf{T}$, where L is the amount of layers. In other words, under the condition that Eq. (3.8) holds, we are able to make statements about the spectrum of the transition matrix $\mathbf{T}$ only using the layer-aggregated transition matrix $\mathfrak{T}$.

Similar to the Fiedler vector, i.e. the eigenvector corresponding to the second smallest eigenvalue of the Laplacian matrix, here we may use the eigenvector v_2 corresponding to the second largest eigenvalue λ_2 of the transition matrix $\mathbf{T}$. The vector v_2 contains negative and positive entries and sums up to zero. If all individual nodes that belong to the same layer correspond to entries of v_2 with the same sign, we consider the layers of $\mathbf{G}$ partitioned according to v_2, which is also called spectral partitioning or spectral bisection [14, 19]. In this case, according to Corollary 1 in the Appendix, it holds that $\lambda_2(T) = \lambda_2(\mathfrak{T})$.

We note that the multi-layer aggregation, performed according to a spectral partitioning, has similarities to spectral coarse-graining [15]. The multi-layer aggregation presented here decreases the state space as well, but still preserves parts of the spectrum.

The spectral properties introduced in this section are important for our ensemble estimations that follow, since we characterize the diffusion process by its convergence efficiency measured through the second-largest eigenvalue $\lambda_2(\mathbf{T})$ of the supra-transition matrix. However, as outlined before, if Eq. (3.8) holds then this eigenvalue is equal to the second-largest eigenvalue $\lambda_2(\mathfrak{T})$ of the aggregated transition matrix $\mathfrak{T}$. Considering that for the construction of $\mathfrak{T}$ we only used aggregated statistics on the network and not the detailed topologies of the inter-links or any of the intra-links of all single layers, this already provides a hint how we can treat a system in the case of limited information.

3.3   Mean-Field Approximation of Ensemble Properties

With the layer aggregation introduced in the previous section, we are now able to deal with multi-layer network ensembles in case of limited information. In our case, this information concerns knowledge either of the inter-link topology between layers or the intra-link topologies of all single layers. For our purpose we define ensembles based on the inter-link densities and intra-link densities of all single layers, more precisely, by using the amount of nodes, the amount of inter-links across any two layers, and the amount of intra-links of all single layers. The number of nodes in individual layers are represented by the vector $\mathbf{n} = \{n_1, \ldots, n_L\}$ and the number of links between layers by a matrix $\mathbf{M}$ with entries m_{st} where s gives the source layer and t the target layer. Intra-layer links have both of their ends in the same layer and therefore we assume that the diagonal elements m_{ss} are equal to the amount of desired intra-links multiplied by two. We denote the ensemble defined by these two quantities $\mathcal{E}(\mathbf{n}, \mathbf{M})$.

A single random realization of this ensemble satisfies the aggregated statistics given by $\mathbf{M}$ and $\mathbf{n}$. We assume a random uniform distribution of links and therefore each realization of $\mathcal{E}(\mathbf{n}, \mathbf{M})$ has the same probability. However, instead of single realizations we are rather interested in the average values of all possible realizations. For each multi-layer network realization $\mathbf{G}$ of $\mathcal{E}(\mathbf{n}, \mathbf{M})$ we build the supra-transition matrix $\mathbf{T}$, which defines a random walk process that is different for every realization. As discussed above, a proxy of the convergence quality of these random walk processes is given by the second-largest eigenvalue $\lambda_2(\mathbf{T})$. Our goal is to estimate the average λ_2 of the ensemble $\mathcal{E}(\mathbf{n}, \mathbf{M})$, and we do this using a mean-field approach on the supra-transition matrix $\mathbf{T}$ that is similar to Refs. [16, 17].

Hereafter we will provide a mean-field approach for the general case, i.e. when the exact topology of inter-links and intra-links of all single layers are unknown. Next, building on this approach, we will discuss the case for which we have full knowledge of the intra-link topology but we do not know the inter-link topology, and the case for which we have full knowledge of the inter-link topology but we do not know the intra-link topology.

Case I: Unknown Topology of Inter-links and Intra-links for All Layers

For this case we only assume knowledge of the ensemble parameters $\mathbf{M}$ and $\mathbf{n}$. We define a mean-field adjacency matrix $\hat{\mathbf{A}}$ with a block structure similar to Eq. (3.4), and for each $\hat{\mathbf{A}}_{st}$ we are only given the amount of links equal to m_{st}. Since we do not know how these links are assigned to the entries $\mathbf{A}_{st}$, without loss of generality we assume a uniform distribution. Thus, for the blocks of $\hat{\mathbf{A}}$ we have

$$\hat{\mathbf{A}}_{st} = \left\{ \frac{m_{st}}{n_s n_t} \right\}_{ij}, \quad i \in \{1, \ldots, n_s\}, \quad j \in \{1, \ldots, n_t\}. \tag{3.11}$$

Following the discussion of Sect. 3.2, based on the mean-field adjacency matrix we define a mean-field transition matrix $\hat{\mathbf{T}}$. The transition probability between any two nodes $i, j \in G_s$ for a fixed layer s is the same since according to the available information individual nodes cannot be distinguished based on their connectivity. Further, the transition probabilities between any two nodes $i \in G_s$ and $j \in G_t$ are the same for any two fixed layers s and t. Therefore, all block transition matrices $\hat{\mathbf{T}}_{st}$ contain the same value at each entry. Hence we have

$$\hat{\mathbf{T}}_{st} = \left\{ \frac{m_{st}}{n_t \left(\sum_k m_{sk} \right)} \right\}_{ij}, \quad i \in \{1, \ldots, n_s\}, \quad j \in \{1, \ldots, n_t\}. \tag{3.12}$$

Now, using Eq. (3.8) we can construct an aggregated supra-transition matrix $\mathfrak{T}$ with entries

$$\alpha_{st} = \frac{m_{st}}{\sum_k m_{sk}}. \tag{3.13}$$

The aggregated supra-transition matrix $\mathfrak{T}$ describes the macro behavior of the multi-layer network ignoring the detailed topology of the inter-links and the intra-links of all single layers. Since $\mathfrak{T}$ depends on a mean-field approach it only captures probabilistic assumptions of the ensemble $\mathcal{E}(\mathbf{n}, \mathbf{M})$. Thus, the spectrum of the mean-field supra-transition matrix $\hat{\mathbf{T}}$ can be calculated by

$$Spec(\hat{\mathbf{T}}) = Spec(\mathfrak{T}) \cup \left(\bigcup_{s=1}^{L} \cup_{i=1}^{n_s-1} \{0\} \right). \tag{3.14}$$

To clarify the situation, let us briefly discuss the simple case of a network $\mathbf{G}$ that contains only two layers G_1 and G_2, for which we get

$$\mathfrak{T} = \begin{pmatrix} 1 - \alpha_{12} & \alpha_{12} \\ \alpha_{21} & 1 - \alpha_{21} \end{pmatrix}. \tag{3.15}$$

Hence, for the mean-field matrix of a two-layered network we obtain

$$Spec(\hat{\mathbf{T}}) = \{1, 1 - \alpha_{12} - \alpha_{21}, \underbrace{0, \ldots, 0}_{|n|-2 \text{ times}}\}. \tag{3.16}$$

These results are remarkable, since the layer-aggregated transition matrix captures the same relevant eigenvalues as the mean-field transition matrix. So, for the case of a diffusion process in two layers the eigenvalue of interest is $\lambda_2(\hat{\mathbf{T}}) = 1 - \alpha_{12} - \alpha_{21}$. However, so far we only considered the general case where we can only use the densities of inter-links and intra-links of all single layers. In the following two sections we will investigate cases where we may have some additional information about either the inter-link topology between all single layers or the intra-link topology of all single layers. For simplicity, we will restrict ourselves to the two layer case but, as shown in the appendix, our results can be generalized to multiple layers.

Case II: Unknown Inter-connectivity

For this case we assume full knowledge of the intra-link topology, i.e. we know exactly which nodes are connected in all of the single layers. But while we know the number of links between the layers we do not know how the layers are connected, i.e. we do not know the inter-link topology. With respect to the general case discussed previously, here we have more information which is expected to improve the estimates of the ensemble average.

More precisely, we consider a two-layer network with unknown inter-link structure denoted by $E_I(\mathbf{G})$, but with a given amount of m interconnecting links which connect the networks G_1 and G_2. This means that the diagonal blocks $\mathbf{A}_1$ and $\mathbf{A}_2$ of the supra-adjacency matrix are given, but the off-diagonal blocks $\mathbf{A}_{12}$ and $\mathbf{A}_{21}$ can take any form such that they have exactly m entries different from zero. Since there are no further constraints on the ensemble, any random link configuration that consists of m inter-links has the same probability to occur. Therefore, we define the mean-field supra-adjacency blocks that correspond to the inter-links, $\hat{\mathbf{A}}_{12}$ and $\hat{\mathbf{A}}_{21}$, to have the same value $\frac{m}{n_1 n_2}$ in each entry.

For the supra-transition matrix we have to row-normalize $\mathbf{A}_1$ with $\hat{\mathbf{A}}_{12}$ and $\hat{\mathbf{A}}_{21}$ with $\mathbf{A}_2$. The row sums of $\hat{\mathbf{A}}_{12}$ are all equal to m/n_1 and the row sums of $\hat{\mathbf{A}}_{21}$ are all equal to m/n_2, while the row sums of $\mathbf{A}_1$ and $\mathbf{A}_2$ correspond to the individual degrees of the nodes in G_1 and G_2 respectively. Thus, we use the mean degree $\widehat{d}_1$ of G_1 and $\widehat{d}_2$ of G_2 in order to obtain the row-normalized transition matrix $\hat{\mathbf{T}}$, and to define the following factors

$$\alpha_1 = \frac{n_1 \widehat{d}_1}{n_1 \widehat{d}_1 + m}, \quad \alpha_2 = \frac{n_2 \widehat{d}_2}{n_2 \widehat{d}_2 + m}, \quad \alpha_{12} = \frac{m}{n_1 \widehat{d}_1 + m}, \quad \alpha_{21} = \frac{m}{n_2 \widehat{d}_2 + m}. \tag{3.17}$$

Note that $\alpha_1 + \alpha_{12} = 1$ and $\alpha_2 + \alpha_{21} = 1$.

Accordingly we define the mean transition blocks of $\mathbf{T}_{12}$ and $\mathbf{T}_{21}$.

$$\hat{\mathbf{T}}_{12} = \left\{ \frac{m}{n_2(n_1 \widehat{d}_1 + m)} \right\}_{ij} \quad \text{for} \quad i \in \{1, \ldots, n_1\}, j \in \{1, \ldots, n_2\} \tag{3.18}$$

$$\hat{\mathbf{T}}_{21} = \left\{ \frac{m}{n_1(n_2 \widehat{d}_2 + m)} \right\}_{ij} \quad \text{for} \quad i \in \{1, \ldots, n_2\}, j \in \{1, \ldots, n_1\}. \tag{3.19}$$

This means that each of the off-diagonal block matrices that correspond to the mean-field inter-link structures have the same value at each matrix element, and the diagonal blocks are just rescaled transition matrices of $\mathbf{A}_1$ and $\mathbf{A}_2$,

$$\hat{\mathbf{T}}_1 = (1 - \alpha_{12}) \, T(\mathbf{A}_1), \quad \hat{\mathbf{T}}_2 = (1 - \alpha_{21}) \, T(\mathbf{A}_2), \tag{3.20}$$

where $T(\mathbf{M})$ is the row-normalized version of matrix $\mathbf{M}$. We denote the supra-transition matrix with the blocks constructed as described before by $\hat{\mathbf{T}}$,

$$\hat{\mathbf{T}} = \left(\begin{array}{c|c} \hat{\mathbf{T}}_1 & \hat{\mathbf{T}}_{12} \\ \hline \hat{\mathbf{T}}_{21} & \hat{\mathbf{T}}_2 \end{array} \right) . \tag{3.21}$$

This mean-field matrix has some special properties. First of all, the eigenvalues of $\hat{\mathbf{T}}_1$ and $\hat{\mathbf{T}}_2$ are also eigenvalues of $\hat{\mathbf{T}}$. Further, the multi-layer aggregation of $\hat{\mathbf{T}}$ is given by

$$\mathfrak{T} = \left(\begin{array}{cc} \alpha_1 & \alpha_{12} \\ \alpha_{21} & \alpha_2 \end{array} \right) = \left(\begin{array}{cc} 1 - \alpha_{12} & \alpha_{12} \\ \alpha_{21} & 1 - \alpha_{21} \end{array} \right) , \tag{3.22}$$

so, the second-largest eigenvalue of $\mathfrak{T}$ is given by $\lambda_2 = 1 - \alpha_{12} - \alpha_{21}$.

The second-largest eigenvalues of $\hat{\mathbf{T}}_1$ is equal to $(1 - \alpha_{12})\lambda_2^1$ and of $\hat{\mathbf{T}}_2$ is equal to $(1 - \alpha_{21})\lambda_2^2$, where $\lambda_2^1 = \lambda_2(T(\mathbf{A}_1))$ and $\lambda_2^2 = \lambda_2(T(\mathbf{A}_2))$. Therefore the second largest eigenvalue of $\hat{\mathbf{T}}$, denoted by $\lambda_2(\hat{\mathbf{T}})$, fulfills the following condition (See Proposition 2 in the Appendix for more details)

$$\lambda_2(\hat{\mathbf{T}}) = \max \left(1 - \alpha_{12} - \alpha_{21}, (1 - \alpha_{12})\lambda_2^1, (1 - \alpha_{21})\lambda_2^2 \right) . \tag{3.23}$$

We would like to remind the reader that an eigenvalue λ_2 close to one implies slow convergence and λ_2 close to zero fast convergence. From the above equation we can see that as long as $\lambda_2 = 1 - \alpha_{12} - \alpha_{21}$ is maximal the inter-links are the limiting factor of the convergence in the multi-layer network. This means that due to the inter-link topology the random walk diffusion is slowed down, and the influence of the intra-layer topologies is marginal to the process.

When either the term of λ_2^1 or λ_2^2 is maximal then the diffusion is limited by the single layer G_1 or G_2, and the additional information provided by the intra-layer topologies becomes relevant as it affects the diffusion process. Note that the change between λ_2 and either λ_2^1 or λ_2^2 being maximal is related to the transitions pointed out in Ref. [3, 18], which is also discussed in Chap. 1.

This behavior is shown in Fig. 3.1 for the mean-field matrix of two interconnected networks. The figure shows the second largest eigenvalues of $\hat{\mathbf{T}}, \mathfrak{T}$ and the sparsest layer $\mathbf{T}_1$ for different amount of inter-links. When only a few inter-links are present the interconnectivity between layers slows the process down, as it is expected. When we increase the amount of inter-links, we can reach the convergence rate of single layers, which is the point where the single layers slow down the process. However,

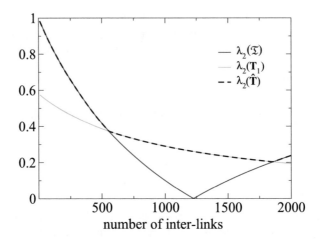

Fig. 3.1 Eigenvalues of a mean-field approach of a two-layered network. Layer 1 consists of an Erdös-Rényi network of 100 nodes and 500 links and Layer 2 consists of an Erdös-Rényi network of 100 nodes and 750 links. The x-axis indicates the amount of inter-links randomly added across the layers. The lines indicate the second-largest eigenvalue of: *black dashed*: the mean-field supra-transition matrix $\lambda_2(\hat{\mathbf{T}})$, *violet*: the layer-aggregated matrix $\lambda_2(\mathfrak{T})$, *turquoise*: the larger single layer eigenvalues of $\lambda_2(\mathbf{T}_1)$ and $\lambda_2(\mathbf{T}_2)$

with an increasing amount of inter-links the single layers lose their importance and the process is again slowed down by the inter-links. This happens because a very large amount of inter-links force the random walker to switch between layers with increasing probability, thus, preventing diffusion to reach the whole layer. To conclude, the mean-field transition matrix $\hat{\mathbf{T}}$ is a better estimation than $\mathfrak{T}$ in intermediate numbers of interlinks, which for our systems is in the region of approximately 550–1800 inter-links. Otherwise, the information about the link densities as captured in $\mathfrak{T}$ is enough to approximate the second-largest eigenvalue of $\hat{\mathbf{T}}$, and thus the speed of diffusion.

In general the spectrum of a mean-field matrix $\hat{\mathbf{T}}$ with unknown inter-link topology is given by

$$Spec(\hat{\mathbf{T}}) = \{1, \lambda_2, \ldots, \lambda_n\} \cup \left(\bigcup_{s=1}^{n} Spec(\hat{\mathbf{T}}_s) \setminus \lambda_1(\hat{\mathbf{T}}_s) \right), \tag{3.24}$$

or

$$Spec(\hat{\mathbf{T}}) = Spec(\mathfrak{T}) \cup \left(\bigcup_{s=1}^{n} Spec(\hat{\mathbf{T}}_s) \setminus \lambda_1(\hat{\mathbf{T}}_s) \right), \tag{3.25}$$

where $\mathfrak{T}$ is the multi-layer aggregation of $\hat{\mathbf{T}}$ as described before (for details see Proposition 2 in the appendix). This decomposition of eigenvalues can also be useful for other network properties that depend on eigenvalues.

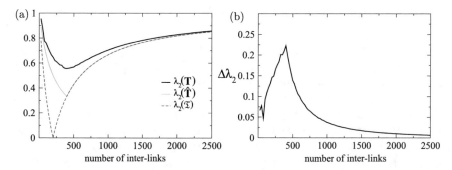

Fig. 3.2 (**a**) Eigenvalues of a mean-field approach of a two-layered network. Layer 1 and 2 both consist of an Erdös-Rényi network of 50 nodes and 100 links but with different topologies. The *x*-axis indicates the amount of random inter-links added across layers. The lines indicate the second-largest eigenvalues of: *black line*: ensemble averages, *turquoise line*: mean-field estimate including intra-link topology, *violet dashed line*: mean-field only considering densities. (**b**) Eigenvalue difference between ensemble average and mean-field estimation $\Delta\lambda_2 = \lambda_2(\mathbf{T}) - \lambda_2(\hat{\mathbf{T}})$

So far we provided an estimation based on the eigenvalues of a mean-field transition matrix $\hat{\mathbf{T}}$ that intends to approximate the ensemble average. In reality however, ensemble realizations of multi-layer networks that contain layer G_1 and G_2 can deviate from the mean-field estimation. This is shown in Fig. 3.2a where we plot the second-largest eigenvalues of $\hat{\mathbf{T}}$, $\mathfrak{T}$, and ensemble averages over 100 realizations of $\mathbf{T}$ against the number of inter-links between G_1 and G_2. As we can see, the magenta colored dashed line showing the mean-field approximation of $\mathfrak{T}$ is a good proxy for the diffusion dynamics in the region when inter-links dominate, which is the case for either sparse or very dense inter-link topologies. However, as shown by the cyan colored line, we can actually improve this approximation if we additionally consider the intra-links of all single layers.

There is a peak where the difference between the estimation and the ensemble averages $\Delta\lambda_2 = \lambda_2(\mathbf{T}) - \lambda_2(\hat{\mathbf{T}})$ reaches high values up to 0.225, as shown in Fig. 3.2b. This happens, on one hand, due to the large degree of freedom that comes from the absence of intra-connectivity informations within the layers. On the other hand, the mean-field matrix assumes "full-connectivity" across layers, and even though this implies small weights for each single inter-link, it leads to a systematic bias towards overestimating the diffusion speed. Nevertheless, we would like to highlight that the multi-layer aggregation provides a quite accurate estimation of the diffusion speed in the regimes where inter-links limit diffusion.

Case III: Unknown Intra-connectivity

For this case we assume full knowledge of the inter-link topology, i.e. we know exactly how the layers are connected, but the intra-link topologies, i.e. how the nodes are connected within the single layers, are unknown. More precisely, we consider two interconnected layers G_1 and G_2 of a multi-layer network, and we fix the inter-links $E_I(\mathbf{G})$ in a bipartite network structure that connects nodes of G_1 to nodes of G_2. Since we have no information about the intra-link topologies of G_1 and G_2, we assume random connectivities within the layers, so that we only know the average degrees $\widehat{d}_1$ and $\widehat{d}_2$ of G_1 and G_2 respectively. This means that the off-diagonal blocks $\mathbf{A}_{12}^\top = \mathbf{A}_{21}$ of the supra-adjacency matrix are given, but the diagonal blocks $\mathbf{A}_1$ and $\mathbf{A}_2$ are unknown.

Because we only know the average degrees $\widehat{d}_1$ and $\widehat{d}_2$ of the layers, we can define mean-field versions of the adjacency matrices such that

$$\widehat{\mathbf{A}_1} = \left\{ \frac{\widehat{d}_1}{n_1} \right\}_{ij} \quad \text{and} \quad \widehat{\mathbf{A}_2} = \left\{ \frac{\widehat{d}_2}{n_2} \right\}_{ij} .$$

However, even though we know the topology of the inter-links, we do not know which nodes exactly are connected to each other. Hence we use the same approach as in Case II with m equal to the amount of inter-links and the factors defined as in Eq. (3.17). Therefore we get the mean-field transition matrix $\widehat{\mathbf{T}}$ consisting of the following block matrices,

$$\widehat{\mathbf{T}}_1 = \left\{ \frac{\widehat{d}_1}{n_1 \widehat{d}_1 + m} \right\}_{ij} \quad \text{for} \quad i \in \{1,\ldots,n_1\}, j \in \{1,\ldots,n_2\} \tag{3.26}$$

$$\widehat{\mathbf{T}}_2 = \left\{ \frac{\widehat{d}_2}{n_2 \widehat{d}_1 + m} \right\}_{ij} \quad \text{for} \quad i \in \{1,\ldots,n_2\}, j \in \{1,\ldots,n_1\} . \tag{3.27}$$

The off-diagonal blocks are just rescaled transition matrices of $\mathbf{A}_{12}$ and $\mathbf{A}_{21}$,

$$\widehat{\mathbf{T}}_{12} = \alpha_{12} T(\mathbf{A}_{12}), \quad \widehat{\mathbf{T}}_{21} = \alpha_{21} T(\mathbf{A}_{21}) . \tag{3.28}$$

However, this time we are not able to compute exactly the single layer eigenvalues λ_1^1 and λ_2^2, as it was the case in Case II. In particular, depending on the ensemble constraints we could only compute an average eigenvalue $\widehat{\lambda}_2$ for a single layer. Therefore, we can use the following maximization term

$$\lambda_2(\widehat{\mathbf{T}}) = \max \left(1 - \alpha_{12} - \alpha_{21}, (1 - \alpha_{12})\widehat{\lambda}_2^1, (1 - \alpha_{21})\widehat{\lambda}_2^2 \right) , \tag{3.29}$$

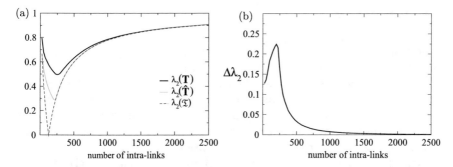

Fig. 3.3 (**a**) Eigenvalues of a mean-field approach of a tow-layered network with 250 inter-links. Layer 1 and 2 both consist of 50 nodes but no edges. The *x*-axis indicates the amount of links intra-links that are simultaneously added to both layers. The lines indicate the second-largest eigenvalue of: *black line*: ensemble averages, *turquoise line*: mean-field estimate including inter-links, *dashed violet line*: mean-field only considering link densities. (**b**) Eigenvalue Difference between ensemble average and mean-field approach $\Delta\lambda_2 = \lambda_2(\mathbf{T}) - \lambda_2(\hat{\mathbf{T}})$

which is has the same form as in Case II (see Eq. (3.23)). Here again, as long as $\lambda_2 = 1 - \alpha_{12} - \alpha_{21}$ is maximal the inter-links are the limiting factor of diffusion in the multi-layer network, which means that due to the inter-link topology the random walk diffusion is slowed down, and the influence of the intra-layer topologies is marginal to the process. On the other hand, when either the average term of $\widehat{\lambda_2^1}$ or $\widehat{\lambda_2^2}$ is maximal then the diffusion is limited by the single layer G_1 or G_2, and the additional information provided by the intra-layer topologies becomes relevant as it affects the diffusion process.

In Fig. 3.3a, starting with initially empty intra-networks,[1] we plot the second largest eigenvalues of $\mathfrak{T}$, $\hat{\mathbf{T}}$, and the ensemble average of 100 realizations of $\mathbf{T}$ against the number of intra-links that are simultaneously and randomly added in both layers. We observe that the general behavior is similar to Fig. 3.2. Thus, the multi-layer aggregation plotted in magenta approximates well the regions where the inter-links are the relevant factor, which is for very sparse and increasingly dense intra-links densities. The difference between the mean-field and the ensemble average $\Delta\lambda_2 = \lambda_2(\mathbf{T}) - \lambda_2(\hat{\mathbf{T}})$ as seen in Fig. 3.3b again rises up to a peak of about 0.225.

Our analysis shows that there is some form of symmetry in knowing the degree of the nodes in the single layers, but not knowing how they are connected to nodes in other layers and to knowing the inter-links between layers, but not the degree of their adjacent nodes. Even though the ensembles generated from these two constraints can be much different, the relevance of inter-links or intra-links of all single layers to a diffusive process is comparable for both cases.

---

[1]Note that even though the intra-layer networks are empty initially, there is a number of inter-layer links which provide connectivity across the layers, similar to a bipartite network.

3.4   Conclusion

In this chapter, we showed how an ensemble perspective can be applied to multi-layer networks in order to address realistic scenarios when only limited information is available. More precisely, we focused on a diffusion process that runs on the multi-layer network and its relation to the spectrum of the supra-transition matrix. We have shown that the convergence rate of the diffusion process is limited by either the inter-links or intra-links of the single layers and we identified for which relation of inter-link compared to intra-link densities it is sufficient to only consider transitions across layers, instead of the full information on all individual nodes. This implies that we do not always need perfect information to make statements about a multi-layer network because, under certain conditions, we are still able to make analytical statements about the network only using partial information. In realistic situations data can be an issue either due to availability constraints or due to their vast amounts. In such cases, even though an exact analysis is impossible, we may still derive useful conclusions about processes that depend on the network spectrum (like diffusion and synchronization) using only aggregated statistics.

For our study we assumed the simplest case of random networks, therefore exploring other ways to couple the network layers or including link-weights and directed links and testing their influence on our results is up to future investigation.

Acknowledgements N.W., A.G. and F.S. acknowledge support from the EU-FET project MUL-TIPLEX 317532.

Appendix

Note: Unless stated otherwise, here vectors are considered to be row-vectors and multiplication of vectors with matrices are left multiplications.

We assume a multi-layer network $\mathbf{G}$ consisting of L layers $G_1, \ldots, G_L$ and n nodes. A single layer G_s contains n_s nodes and therefore $\sum_{s=1}^{L} n_s = n$. For a multi-layer network $\mathbf{G}$ we define the *supra-transition* matrix that can be represented in block structure according to the layers:

$$
\mathbf{T} = \begin{pmatrix}
\mathbf{T}_1 & \cdots & \mathbf{T}_{1t} & \cdots & \mathbf{T}_{sL} \\
\vdots & \ddots & \vdots & \ddots & \vdots \\
\mathbf{T}_{s1} & \cdots & \mathbf{T}_{st} & \cdots & \mathbf{T}_{sL} \\
\vdots & \ddots & \vdots & \ddots & \vdots \\
\mathbf{T}_{L1} & \cdots & \mathbf{T}_{Lt} & \cdots & \mathbf{T}_L
\end{pmatrix}.
$$

Each $\mathbf{T}_{st}$ contains all the transition probabilities from nodes in G_s to nodes in G_t. Assuming Eq. (3.8) it follows that $\mathbf{T}_{st} = \alpha_{st}\mathbf{R}_{st}$ where $\mathbf{R}_{st}$ is a row stochastic matrix.

This means that all $\mathbf{T}_{st}$ are scaled transition matrices. The factor α_{st} represents the weighted fraction of all links starting in G_s that end up in G_t.

In this respect we define the aggregated transition matrix $\mathfrak{T}$ of dimension L,

$$
\mathfrak{T} = \begin{pmatrix}
\alpha_{11} & \cdots & \alpha_{1t} & \cdots & \alpha_{sL} \\
\vdots & \ddots & \vdots & \ddots & \vdots \\
\alpha_{s1} & \cdots & \alpha_{st} & \cdots & \alpha_{sL} \\
\vdots & \ddots & \vdots & \ddots & \vdots \\
\alpha_{L1} & \cdots & \alpha_{Lt} & \cdots & \alpha_{L}
\end{pmatrix} .
$$

Each vector v of dimension n can be split according to the layer-separation given by $\mathbf{G}$,

$$
v = \left(v^{(1)}, \ldots, v^{(k)}, \ldots, v^{(L)} \right) .
$$

Each component $v^{(k)}$ has exactly dimension n_k. We define the *layer-aggregated vector* $\mathfrak{v} = (\mathfrak{v}_1, \ldots, \mathfrak{v}_L)$ of dimension L as follows

$$
\forall k \in \{1, \ldots, L\} \quad \mathfrak{v}_k = \sum_{i=1}^{n_k} \left[v^{(k)} \right]_i .
$$

We use the bracket notation $[v]_i$ to represent the i-th entry of the vector v. Analogously, by $[v\mathbf{M}]_i$ we mean the i-th entry w_i that represents the multiplication of v with a matrix $\mathbf{M}$, i.e. $w = v\mathbf{M}$. Further, by $|v|$ we indicate the sum of the entries of v, $|v| = \sum_i v_i = \sum_i [v]_i$.

Theorem 1 *For a multi-layer network $\mathbf{G}$ consisting of L layers we assume the supra-transition matrix $\mathbf{T}$ to consist of block matrices $\mathbf{T}_{st}$ such that for all $s, t \in \{1, \ldots, L\}$, $\mathbf{T}_{st} = \alpha_{st} \mathbf{R}_{st}$ where $\alpha_{st} \in \mathbb{Q}$ and $\mathbf{R}_{st}$ is a row stochastic transition matrix. The multi-layer aggregation is defined by $\mathfrak{T} = \{\alpha_{st}\}_{st}$. If an eigenvalue λ of the matrix $\mathbf{T}$ corresponds to an eigenvector v with a layer-aggregation $\mathfrak{v}$ that satisfies $\mathfrak{v} \neq 0$ then λ is also an eigenvalue of $\mathfrak{T}$.*

Proof Assume v is a left eigenvector of $\mathbf{T}$ corresponding to the eigenvalue λ. Therefore, it holds that $v\mathbf{T} = \lambda v$. Let $v^{(k)}$ be the k-th part of v that corresponds to the layer G_k. We can write the matrix multiplication in block structure.

$$
\left(v^{(1)}, \ldots, v^{(k)}, \ldots, v^{(L)} \right) \mathbf{T} = \left(\sum_l v^{(l)} \mathbf{T}_{l1}, \ldots, \sum_l v^{(l)} \mathbf{T}_{lk}, \ldots, \sum_l v^{(l)} \mathbf{T}_{lL} \right) .
$$

Each $v^{(k)}$ is a row vector which length is equal to the amount of nodes n_k in G_k. The transformation $\sum_l v^{(l)} \mathbf{T}_{lk}$ is also a row vector with the same length as $v^{(k)}$.

According to the eigenvalue equation it holds that for all $k \in \{1, \ldots, L\}$

$$\lambda v^{(k)} = (v\mathbf{T})^{(k)} = \sum_l v^{(l)} \mathbf{T}_{lk} \, .$$

Now let us denote the sum of the vector entries of $v^{(k)}$ as

$$\mathfrak{v}_k = \sum_i \left[v^{(k)} \right]_i \, .$$

Further, we define layer-aggregated vector consisting of this sums by $\mathfrak{v} = (\mathfrak{v}_1, \ldots, \mathfrak{v}_n)$. Note that for a general row stochastic matrix $\mathbf{M}$ and its multiplication with any vector v it holds that $\sum_j [v]_j = \sum_j [v\mathbf{M}]_j$. For the components after multiplication with $\mathbf{T}$ we can deduce

$$\sum_i \left[(v\mathbf{T})^{(k)} \right]_i = \sum_i \left[\sum_l v^{(l)} \mathbf{T}_{lk} \right]_i = \sum_i \left[\sum_l \alpha_{lk} v^{(l)} \mathbf{R}_{lk} \right]_i$$

$$= \sum_l \alpha_{lk} \sum_i \left[v^{(l)} \mathbf{R}_{lk} \right]_i = \sum_l \alpha_{lk} \sum_i \left[v^{(l)} \right]_i = \sum_l \alpha_{lk} \mathfrak{v}_l \, .$$

If we multiply $\mathfrak{v}$ with $\mathfrak{T}$ and look at a single entry of $\mathfrak{v}\mathfrak{T}$ we get

$$[\mathfrak{v}\mathfrak{T}]_k = \sum_l \mathfrak{v}_l \mathfrak{T}_{lk} = \sum_l \mathfrak{v}_l \alpha_{lk} \, .$$

Hence it holds that

$$[\mathfrak{v}\mathfrak{T}]_k = \sum_i \left[(v\mathbf{T})^{(k)} \right]_i \, ,$$

and therefore

$$\mathfrak{v}\mathfrak{T} = \left(\sum_i \left[(v\mathbf{T})^{(1)} \right]_i, \ldots, \sum_i \left[(v\mathbf{T})^{(L)} \right]_i \right) \, .$$

Finally since $\mathbf{T}$ is row stochastic and $\lambda v = v\mathbf{T}$ we have that

$$\lambda \mathfrak{v} = \lambda \, (\mathfrak{v}_1, \ldots, \mathfrak{v}_n)$$

$$= \lambda \left(\sum_i [v^{(1)}]_i, \ldots, \sum_i [v^{(L)}]_i \right)$$

$$= \left(\sum_i \left[(v\mathbf{T})^{(1)} \right]_i, \ldots, \sum_i \left[(v\mathbf{T})^{(L)} \right]_i \right)$$

$$= \mathfrak{v}\mathfrak{T} \, .$$

Therefore, λ is also an eigenvalue of $\mathfrak{T}$ to the eigenvector $\mathfrak{v}$ defined as before. It is important to note that this only holds if $\mathfrak{v} \neq 0$.

The procedure used in the proof of the previous theorem applies to several eigenvalues of $\mathbf{T}$ but at most L of them. Next we give a proposition for the reversed statement of Theorem 1.

Proposition 1 *Let* $\mathbf{G}$ *be a multi-layer network that consists of* L *layers and fulfills all of the conditions of Theorem 1. Let* $\mathfrak{T} = \{\alpha_{st}\}_{st}$ *be the multi-layer aggregation of* $\mathbf{T}$. *If* λ *is an eigenvalue of* $\mathfrak{T}$ *then* λ *is also an eigenvalue of* $\mathbf{T}$.

Proof Assume that λ is an eigenvalue of $\mathfrak{T}$. For each matrix there exist a left and right eigenvector that correspond to the same eigenvalue λ. Assume the $\mathfrak{w}$ is the right eigenvector and therefore a column vector. Hence $\mathfrak{T}\mathfrak{w} = \lambda\mathfrak{w}$ and

$$\mathfrak{T}\mathfrak{w} = \left(\sum_j \alpha_{1j}\mathfrak{w}_j, \ldots, \sum_j \alpha_{kj}\mathfrak{w}_j, \ldots, \sum_j \alpha_{Lj}\mathfrak{w}_j \right)^{\mathsf{T}} = \lambda\mathfrak{w}.$$

Now we generate a column vector w of dimension n such that for all the layer components $w^{(k)}$ it holds that

$$w^{(k)} = (\mathfrak{w}_k, \ldots, \mathfrak{w}_k)^{\mathsf{T}}.$$

Next we perform a right multiplication of w with $\mathbf{T}$,

$$\mathbf{T}w = \left(\sum_l \mathbf{T}_{1l}w^{(l)}, \ldots, \sum_l \mathbf{T}_{kl}w^{(l)}, \ldots, \sum_l \mathbf{T}_{Ll}w^{(l)} \right)^{\mathsf{T}}$$

$$= \left(\sum_l \alpha_{1l}\mathbf{R}_{1l}w^{(l)}, \ldots, \sum_l \alpha_{kl}\mathbf{R}_{kl}w^{(l)}, \ldots, \sum_l \alpha_{Ll}\mathbf{R}_{Ll}w^{(l)} \right)^{\mathsf{T}}.$$

Since all $\mathbf{R}_{st}$ are row stochastic matrices and $w^{(l)}$ contains only the value $\mathfrak{w}_l$ for each entry we get $\mathbf{R}_{st}w^{(l)} = w^{(l)}$. It follows that

$$\mathbf{T}w = \left(\sum_l \alpha_{1l}w^{(l)}, \ldots, \sum_l \alpha_{kl}w^{(l)}, \ldots, \sum_l \alpha_{Ll}w^{(l)} \right)^{\mathsf{T}}$$

$$= \left(\lambda w^{(1)}, \ldots, \lambda w^{(k)}, \ldots, \lambda w^{(L)} \right)^{\mathsf{T}}$$

$$= \lambda w.$$

And therefore λ is also an eigenvalue of $\mathbf{T}$.

In the case of a diffusion process we are especially interested in the second-largest eigenvalue of $\mathbf{T}$, denoted by $\lambda_2(\mathbf{T})$, which is related to algebraic connectivity of $\mathbf{T}$. In this perspective the following corollary is useful:

Corollary 1 *Let* G *be a multi-layer network consisting of* L *layers that fulfill all of the conditions of Theorem 1. Further assume that* $\mathbf{G}$ *is partitioned according to a spectral partitioning, i.e. according to the eigenvector corresponding to* $\lambda_2(\mathbf{T})$*, then* $\lambda_2(\mathbf{T}) = \lambda_2(\mathfrak{T})$.

Proof In general all the eigenvectors of a transition matrix, except the eigenvector corresponding to the largest eigenvalue that is equal to one, sum up to zero. However, these eigenvectors consist of positive and negative entries that allow for a spectral partitioning. Especially the eigenvector v_2 that corresponds to the second-largest eigenvalue $\lambda_2(\mathbf{T})$, can be used for the partitioning of the network. This eigenvector is related to the Fiedler vector that is also used for spectral bisection [19]. Therefore if the layer-partition of $\mathbf{G}$ coincides with this spectral partitioning we assure that the layer-aggregated vector of v_2 satisfies $\mathfrak{v}_2 \neq 0$. Considering this and Proposition 1 the corollary follows directly from Theorem 1.

Given Eq. (3.8) we can fully describe the spectrum of $\mathbf{T}$ based on the intra-layers transition blocks $\mathbf{T}_i$ for $i \in \{1, \ldots, n\}$ and the spectrum of $\mathfrak{T}$. Note that with uniform columns of a matrix $\mathbf{M}$ we mean that each column of $\mathbf{M}$ contains the same value in each entry. However, this value can be different for different columns.

Proposition 2 *Let* $\mathbf{T}$ *be the supra-transition matrix of a multi-layer network* $\mathbf{G}$ *that consist of* L *layers and satisfies Eq. (3.8). If* $\mathbf{T}$ *has off-diagonal block matrices* $\mathbf{T}_{st}$*, for* $s, t \in \{1, \ldots, n\}$ *and* $s \neq t$*, that all have uniform columns, then the spectrum of* $\mathbf{T}$ *can be decomposed as*

$$Spec(\mathbf{T}) = \{1, \lambda_2, \ldots, \lambda_L\} \cup \left(\bigcup_{s=1}^{L} Spec(\mathbf{T}_s) \setminus \{\lambda_1(\mathbf{T}_s)\} \right),$$

where $\mathbf{T}_s$ *are the block matrices of* $\mathbf{T}$ *corresponding to the single layers* G_s *and* $\lambda_1(\mathbf{T}_s)$ *the largest eigenvalue of* $\mathbf{T}_s$*. The eigenvalues* $\lambda_2, \ldots, \lambda_L$ *are attributed to the interconnectivity of layers.*

Proof To prove this statement we just have to show that all eigenvalues (except the largest one) of $\mathbf{T}_s$ for $s \in \{1, \ldots, L\}$ are also eigenvalues of $\mathbf{T}$. We assume that λ is any eigenvalue corresponding to the eigenvector u of some block matrix $\mathbf{T}_r$, i.e. $\lambda u = u\mathbf{T}_r$. We define a row vector v that is zero everywhere except at the position where it corresponds to $\mathbf{T}_r$. The vector v looks like $v = (0, \ldots, 0, u, 0, \ldots, 0)$. Now we investigate what happens if we multiply this vector with the transition matrix $\mathbf{T}$.

$$v\mathbf{T} = \left(v^{(1)}, \ldots, v^{(k)}, \ldots, v^{(L)} \right) \mathbf{T} = \left(\sum_l v^{(l)} \mathbf{T}_{l1}, \ldots, \sum_l v^{(l)} \mathbf{T}_{lk}, \ldots, \sum_l v^{(l)} \mathbf{T}_{lL} \right).$$

Let us take a look at the effect of the matrix multiplication on an arbitrary component $v^{(k)}$ with $k \neq r$ and recall that $v^{(k)}$ is equal to a zero vector $\mathbf{0}$ for $k \neq r$.

$$(v\mathbf{T})^{(k)} = \sum_l v^{(l)}\mathbf{T}_{lk} = \sum_{l,l\neq r} \mathbf{0}\mathbf{T}_{lk} + u\mathbf{T}_{rk} = u\mathbf{T}_{rk}.$$

Note that all eigenvectors u of a transition matrix that are not related to the largest eigenvalue sum up to zero. Therefore it holds that $u\mathbf{T}_{rk} = \mathbf{0}$ since $\mathbf{T}_{rk}$ has uniform columns and therefore $u\mathbf{T}_{rk}$ yields in a vector where each entry is equal to some multiple of $|u|$. In case of $k = r$ it holds that $v^{(k)} = u$ and we get

$$(v\mathbf{T})^{(r)} = \sum_l v^{(l)}\mathbf{T}_{lr} = \sum_{l,l\neq r} \mathbf{0}\mathbf{T}_{lk} + u\mathbf{T}_{rr} = u\mathbf{T}_r = \lambda r.$$

Hence, it holds that $v\mathbf{T} = \lambda v$, which means that λ is also an eigenvalue of $\mathbf{T}$. This way we get $n - L$ eigenvalues of $\mathbf{T}$ apart from the largest eigenvalue that is equal to one. The remaining eigenvalues denoted by $\lambda_2, \ldots, \lambda_L$ are not attributed to any block matrix of $\mathbf{T}$. Therefore they are considered to be the interconnectivity eigenvalues.

Corollary 2 *Let* $\mathbf{G}$ *be a multi-layer network consisting of L layers that satisfies Eq. (3.8) and the conditions of Proposition 2. Then the aggregated matrix* $\mathfrak{T} = \{\alpha_{st}\}_{st}$ *has spectrum*

$$Spec(\mathfrak{T}) = \{1, \lambda_2, \ldots, \lambda_L\},$$

and it holds that $\lambda_2, \ldots, \lambda_L \in Spec(\mathbf{T})$.

Proof Note that every eigenvalue $\lambda \neq 1$ of some block matrix $\mathbf{T}_r$ with $\lambda u = u\mathbf{T}_r$ is by Proposition 2 also an eigenvalue of $\mathbf{T}$. Furthermore, λ is attributed to the eigenvector $v = (0, \ldots, 0, u, 0, \ldots, 0)$ of $\mathbf{T}$. However $|v| = 0$ since u is an eigenvector of a transition matrix, not related to the largest eigenvalue, and therefore sums up to zero. Hence all eigenvalues fulfilling this condition are by Theorem 1 not eigenvalues of $\mathfrak{T}$. Since $\mathfrak{T}$ contains at least L eigenvalues that by Proposition 1 also correspond to eigenvalues of $\mathbf{T}$, the remaining eigenvalues $\lambda_2, \ldots, \lambda_L$ have to also be eigenvalues of $\mathfrak{T}$.

In the following we provide a useful proposition for the eigenvalues arising from the inter-links in case of two layers. Note that by the function $T(\cdot)$ applied to matrix $\mathbf{M}$ we indicate that $T(\mathbf{M})$ is the row-normalization of $\mathbf{M}$.

Proposition 3 *Let* $\mathbf{G}$ *be a multi-layer network that satisfies Eq. (3.8), consisting of two networks* G_1 *and* G_2 *in separate layers. Assume that the supra-transition matrix* $\mathbf{T}$ *has the form*

$$\mathbf{T} = \left(\begin{array}{c|c} \mathbf{T}_1 & \mathbf{T}_{12} \\ \hline \mathbf{T}_{21} & \mathbf{T}_2 \end{array} \right) = \left(\begin{array}{c|c} \beta T(\mathbf{A}_1) & (1-\beta)\mathbf{T}'_{12} \\ \hline (1-\beta)\mathbf{T}'_{21} & \beta T(\mathbf{A}_2) \end{array} \right),$$

where $\mathbf{T}^I$ is the transition matrix of the layer $\mathbf{G}$ that only consists of the inter-layer links and $\beta \in \mathbb{Q}$ is a constant. Furthermore, assume that $\mathbf{T}_1$ and $\mathbf{T}_2$ have uniform columns.

Then for $\lambda \in Spec(\mathbf{T}^I)$ and $\lambda \neq 1, -1$ it holds that $(1 - \beta)\lambda \in Spec(\mathbf{T})$.

Proof If v is an eigenvector to the eigenvalue $\lambda \neq 1, -1$ of $\mathbf{T}^I$ it holds that $v\mathbf{T}^I = \lambda v$. Hence,

$$v\mathbf{T}^I = \left(v^{(1)}, v^{(2)}\right)\mathbf{T}^I = \left(v^{(2)}\mathbf{T}^I_{21}, v^{(1)}\mathbf{T}^I_{12},\right) = \lambda\left(v^{(1)}, v^{(2)}\right).$$

Because $\lambda v^{(2)} = v^{(1)}\mathbf{T}^I_{12}$, we get $\lambda^2 v^{(1)} = v^{(1)}\mathbf{T}^I_{12}\mathbf{T}^I_{21}$. Therefore, $v^{(1)}$ is also an eigenvector of the transition matrix $\mathbf{T}^I_{12}\mathbf{T}^I_{21}$ to the eigenvalue λ^2. Note that $\lambda \neq 1, -1$ hence $\lambda^2 < 1$ which implies that $v^{(1)}$ does not correspond to the largest eigenvalue and therefore its entries sum up to zero. The same holds for $v^{(2)}$ and the matrix $\mathbf{T}^I_{21}\mathbf{T}^I_{12}$. For the multiplication of v with $\mathbf{T}$ we deduce that

$$v\mathbf{T} = \left(v^{(1)}, v^{(2)}\right)\mathbf{T} = \left(v^{(1)}\mathbf{T}_1 + (1 - \beta)v^{(2)}\mathbf{T}^I_{21}, (1 - \beta)v^{(1)}\mathbf{T}^I_{12} + v^{(2)}\mathbf{T}_2,\right).$$

Since $\mathbf{T}_1$ and $\mathbf{T}_2$ have uniform columns we get $v^{(1)}\mathbf{T}_1 = \mathbf{0}$ and $v^{(2)}\mathbf{T}_2 = \mathbf{0}$. And therefore $v\mathbf{T} = (1 - \beta)\lambda v$ and $(1 - \beta)\lambda \in Spec(\mathbf{T})$.

Proposition 3 can be extended to multiple layers, however the proof is more involved and will be omitted.

References

1. Parshani, R., Buldyrev, S.V., Havlin, S.: Critical effect of dependency groups on the function of networks. Proc. Natl. Acad. Sci. **108**, 1007–1010 (2011)
2. Gao, J., Buldyrev, S.V., Stanley, H.E., Havlin, S.: Networks formed from interdependent networks. Nat. Phys. **8**, 40–48 (2012)
3. Gómez, S., Díaz-Guilera, A., Gómez-Gardeñes, J., Pérez-Vicente, C.J.: Moreno, Y., Arenas, A.: Diffusion dynamics on multiplex networks. Phys. Rev. Lett. **110**, 028701 (2013)
4. Garas, A.: Reaction-diffusion processes on interconnected scale-free networks (2014). arXiv:1407.6621
5. Boccaletti, S., Bianconi, G., Criado, R., del Genio, C.I., Gómez-Gardeñes, J., Romance, M., Sendiña-Nadal, I., Wang, Z., Zanin, M.: The structure and dynamics of multilayer networks. Phys. Rep. **544**, 1–122 (2014)
6. Blanchard, P., Volchenkov, D.: Random Walks and Diffusions on Graphs and Databases. Springer, Berlin/Heidelberg (2011). ISBN:978-3-642-19592-1
7. Rosenthal, J.S.: Convergence rates for Markov chains. SIAM Rev. **37**, 387–405 (1995)
8. Chung, F.: Laplacians and the Cheeger inequality for directed graphs. Ann. Comb. **9**, 1–19 (2005)
9. Lovász, L.: Random walks on graphs: a survey. Comb., Paul Erdos Eighty **2**, 1–46 (1993)
10. De Domenico, M., Sole, A., Gómez, S., Arenas, A.: Navigability of interconnected networks under random failures. Proc. Natl. Acad. Sci. **111**, 8351–8356 (2014)
11. Solé-Ribalta, A., De Domenico, M., Kouvaris, N.E., Díaz-Guilera, A., Gómez, S., Arenas, A.: Spectral properties of the Laplacian of multiplex networks. Phys. Rev. E **88**, 032807 (2013)

12. Kivelä, M., Arenas, A., Barthelemy, M., Gleeson, J.P., Moreno, Y., Porter, M.A.: Multilayer networks. J. Complex Netw. **2**, 203–271 (2014)
13. De Domenico, M., Solé-Ribalta, A., Cozzo, E., Kivelä, M., Moreno, Y., Porter, M.A., Gómez, S., Arenas, A.: Mathematical formulation of multilayer networks. Phys. Rev. X **3**, 041022 (2013)
14. Fiedler, M.: A property of eigenvectors of nonnegative symmetric matrices and its application to graph theory. Czechoslov. Math. J. **25**(4), 619–633 (1975)
15. Gfeller, D., De Los Rios, P.: Spectral coarse graining of complex networks. Phys. Rev. Lett. **99**, 038701 (2007)
16. Martín-Hernández, J., Wang, H., Van Mieghem, P., D'Agostino, G.: Algebraic connectivity of interdependent networks. Phys. A: Stat. Mech. Appl. **404**, 92–105 (2014)
17. Grabow, C., Grosskinsky, S., Timme, M.: Small-world network spectra in mean-field theory. Phys. Rev. Lett. **108**: 218701 (2012)
18. Radicchi, F., Arenas, A.: Abrupt transition in the structural formation of interconnected networks. Nat. Phys. **9**, 717–720 (2013)
19. Fiedler, M.: Algebraic connectivity of graphs. Czechoslov. Math. J. **23**, 298–305 (1973)

Chapter 4
Interconnecting Networks: The Role of Connector Links

Javier M. Buldú, Ricardo Sevilla-Escoboza, Jacobo Aguirre, David Papo, and Ricardo Gutiérrez

Abstract Recently, some studies have started to show how global structural properties or dynamical processes such as synchronization, robustness, cooperation, transport or epidemic spreading change dramatically when considering a network of networks, as opposed to networks in isolation. In this chapter we examine the effects that the particular way in which networks get connected exerts on each of the individual networks. We describe how choosing the adequate connector links between networks may promote or hinder different structural and dynamical properties of a particular network. We show that different connecting strategies have consequences on the distribution of network centrality, population dynamics or spreading processes. The importance of designing adequate connection strategies is illustrated with examples of social and biological systems. Finally, we discuss how this new approach can be translated to other dynamical processes, such as synchronization in an ensemble of networks.

J.M. Buldú (✉)
Laboratory of Biological Networks, Center for Biomedical Technology, UPM, Pozuelo de Alarcón, 28223 Madrid, Spain

Complex Systems Group, Universidad Rey Juan Carlos, 28933 Móstoles, Madrid, Spain
e-mail: jmbuldu@gmail.com

R. Sevilla-Escoboza
Centro Universitario de los Lagos, Universidad de Guadalajara, Enrique Díaz de Leon, Paseos de la Montaña, Lagos de Moreno, Jalisco 47460, México
e-mail: sevillaescoboza@gmail.com

J. Aguirre
Centro Nacional de Biotecnología (CSIC). C/ Darwin 3, 28049 Madrid, Spain
e-mail: jaguirre@cnb.csic.es

D. Papo
Laboratory of Biological Networks, Center for Biomedical Technology, UPM, Pozuelo de Alarcón, 28223 Madrid, Spain
e-mail: papodav@gmail.com

R. Gutiérrez
Department of Chemical Physics, The Weizmann Institute of Science, Rehovot 76100, Israel
e-mail: rcd.gutierrez@gmail.com

© Springer International Publishing Switzerland 2016
A. Garas (ed.), *Interconnected Networks*, Understanding Complex Systems,
DOI 10.1007/978-3-319-23947-7_4

4.1 Introduction

During more than a decade, the application of Complex Networks Theory to real systems has given fruitful results in the understanding of how networked systems organize, interact and evolve [1–4]. Initially, the main motivation was to characterize the topology of real systems (randomness, heterogeneity, modularity, etc.) and its connection with structural problems such as resilience, robustness or navigability [5–9]. Then, attention was devoted to how dynamical processes such as synchronization [10, 11], spreading [12–15] or congestion [16–18] were constrained by the network structure [19]. In a further development, the interplay between structure and dynamics was interpreted as a closed loop, wherein the structural properties of networks could be understood as a consequence of an adaptative process influenced by the dynamics and vice versa [20].

More recently, the idea that a network is, in many real cases, a *network of networks* (NoN), has emerged [21, 22]. In many cases, component networks of a NoN can be interpreted as modules of a unique modular network. While the detection and analysis of modules inside a network has been deeply studied [23, 24], the influence of intranetwork structures on dynamical processes remains largely unexplored. For instance, as shown in Ref. [21], interconnections between networks may play a crucial role in processes such as percolation, eventually leading to dramatic first order transitions. Other example is epidemic spreading, where it was shown that the creation of links between the most central nodes of two communities can enhance the propagation of a disease through the whole network [25].

In this chapter we focus on the competition taking place when two initially separated networks are coupled with one or more *connector links* to form a unique *ensemble network*. In particular, we examine how one or both networks can be better off according to some criterion depending on the connecting strategy that is adopted. To determine which network is benefitting the most from the interaction, we make use of the eigenvector centrality [4]. The eigenvector centrality is a measure of node importance that is obtained by calculating the eigenvector associated to the largest eigenvalue of the connectivity matrix, which, as we will see, depends on the dynamical process occurring in the network. Next, the centrality captured by each competing network is obtained as the sum of the centrality of all its nodes. The whole problem can then be framed as a competition for limited resources, since an increase of centrality for one network necessarily entails a corresponding decrease in that of its competitors.

The advantage of such a way of analyzing network competition is that, in addition of being a measure of node importance, the eigenvector centrality is related to a series of dynamical processes, such as disease spreading, diffusion processes, evolution of genotypes, rumor and opinion formation (see Ref. [4] for a review). In these cases, the transient or final state of the system depends directly on the eigenvector $\mathbf{u}_1$ associated to the largest eigenvalue λ_1 of the connectivity matrix.

We will describe how the eigenvector $\mathbf{u}_1$ of two isolated networks is modified when certain connections between them are created, leading to an interconnected network [21, 22].

In the remainder of this chapter we first analyze how the eigenvector $\mathbf{u}_1$ of a NoN can be obtained from the spectral properties of the networks forming the ensemble. We then identify the optimal strategies that a network can follow when connecting to other networks and apply this methodology to population dynamics and epidemic spreading. We finally discuss the main concepts introduced in this chapter and point to possible problems to be tackled in the future.

4.2 The Influence of Interconnectivity on the Spectral Properties of an Interconnected Network

In this section we give analytical expressions for the spectral properties associated to a generic connectivity matrix $\mathbf{M}$, resulting from the connection of two initially isolated networks A and B [26]. The connectivity matrix is a weighted version of the classical adjacency matrix $\mathbf{A}$, where the component M_{ij} measures the strength of the connection between nodes i and j (and $M_{ij} = 0$ if i and j are not connected to each other). The aim is to gain a priori knowledge of the main spectral properties of the interconnected network by inspecting the structure of A and B, and the link(s) connecting both networks. Specifically, we are interested in: (i) the highest eigenvalue of the connectivity matrix and (ii) its associated eigenvector.

Figure 4.1 schematically represents two independent networks A and B, of N_A and N_B nodes and L_A and L_B links respectively, which initially form the disconnected network AB of $N_A + N_B$ nodes and $L_A + L_B$ links. Next, we connect both networks via a set $\{cl\}_{l=1,...,L}$ of L connector links to create a *total interconnected network* T of $N_T = N_A + N_B$ nodes and $L_T = L_A + L_B + L$ links. The adjacency matrix $\mathbf{G_T}$ corresponding to network T is therefore formed by adding to the block diagonal network containing the original adjacency matrices of A and B, $\mathbf{G_{AB}}$, the connector links. For simplicity, let us suppose that $\mathbf{G_T}$ is symmetric, that is, the links of network T are bidirectional (this is tantamount to considering the initially isolated networks A and B to be symmetric and establishing interconnecting links that are bidirectional). Depending on the topological importance of the nodes that act as connectors between networks, four different strategies in the election of a connector link can be adopted: (a) peripheral-peripheral (PP), (b) peripheral-central (PC), (c) central-central (CC) and (d) central-peripheral (CP). Let us call $\lambda_{A,i}$ and $\lambda_{B,i}$ the i eigenvalues of the connectivity matrices $\mathbf{M_A}$ and $\mathbf{M_B}$ respectively, where i goes from 1 to the size of the corresponding network (N_A or N_B) with $i = 1$ corresponding to the largest eigenvalue and the rest following in decreasing order. The relation between connectivity matrices such as $\mathbf{M_A}$, $\mathbf{M_B}$ and $\mathbf{M_T}$ and the adjacency matrices such as $\mathbf{G_A}$, $\mathbf{G_B}$ and $\mathbf{G_T}$ depends on the peculiarities of the process. Let us suppose $\lambda_{A,1} > \lambda_{B,1}$ throughout the chapter, being the *strong* network the one with highest

Strong Network

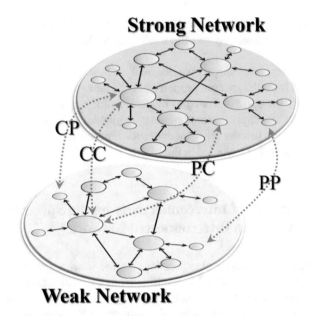

Weak Network

Fig. 4.1 Schematic representation of the different strategies for connecting two networks, according to the centrality of the connector nodes. The strong network is defined as the network with higher λ_1 (first eigenvector of the connectivity matrix **M**). Central nodes C are those with the highest eigenvector centrality, obtained from $\mathbf{u}_1$ (eigenvector associated to λ_1), while peripheral nodes P have the lowest centrality. Initially, the networks remain disconnected and, next, we connect them by adding connector links. According to the centrality of the connector nodes, four different strategies can be followed: (a) peripheral-peripheral (PP), (b) peripheral-central (PC), (c) central-central (CC) and (d) central-peripheral (CP)

λ_1 and the *weak* network the one with the lowest. This way, from now on, network A (B) will be the strong (weak) network.

We call $\mathbf{u}_{A,i}$ the N_A vectors of length N_T where the first N_A elements coincide with the eigenvector i of matrix $\mathbf{M_A}$ and the rest are equal to zero, while $\mathbf{u}_{B,i}$ are the N_B vectors of length N_T where the first N_A elements are zeros and the rest coincide with the eigenvector i of matrix $\mathbf{M_B}$. $\lambda_{T,i}$ and $\mathbf{u}_{T,i}$ are the eigenvalues and eigenvectors of matrix $\mathbf{M_T}$. The main idea of the analytical calculations is to describe the total graph T as a perturbation of graph A by graph B, in a way that the weight of the connector links is $\epsilon \ll 1$. Therefore, as $\lambda_{A,1} > \lambda_{B,1}$ by construction, the maximum eigenvalue $\lambda_{T,1}$ will be a perturbation of $\lambda_{A,1}$ and its associated eigenvector $\mathbf{u}_{T,1}$ will be a perturbation of $\mathbf{u}_{A,1}$. This methodology is inspired by the perturbation theory of matrices presented in [27], and among some other examples it was recently applied in the context of Complex Network Theory to characterize the importance of network nodes and links [28], and for the detection of communities [29]. We give a fully detailed calculation in the Appendix, which ends up with

$$\mathbf{u}_{T,1} = \mathbf{u}_{A,1} + \epsilon \sum_{k=1}^{N_B} a_k \mathbf{u}_{B,k} + o(\epsilon^2), \tag{4.1}$$

where $a_k = (\mathbf{u}_{A,1}\mathbf{P}\mathbf{u}_{B,k})/(\lambda_{A,1} - \lambda_{B,k})$, and $\mathbf{P}$ is a matrix representing the connector links in such a way that $\mathbf{M_T} = \mathbf{M_{AB}} + \epsilon\mathbf{P}$. Note that, since $\mathbf{u}_{A,1}\mathbf{P}\mathbf{u}_{B,1} = \sum (\mathbf{u}_{A,1})_i P_{ij}(\mathbf{u}_{B,1})_j$, only the connector nodes (i.e., those connected by P_{ij}) will contribute to this latter term. See [26] for more details and its application to the case of directed networks (i.e., asymmetric networks, with unidirectional links) and more than two networks.

4.3 Identifying Successful Strategies

The eigenvector $\mathbf{u}_{T,1}$ can be used to determine the outcome of a competition between networks A and B. In this section, we focus on how two networks compete for acquiring the maximum importance inside the interconnected network, while in the next sections we will discuss how to apply these concepts to other dynamical processes. The eigenvector centrality, which is given directly by the eigenvector $\mathbf{u}_{T,1}$ is used as a measure of the topological importance of a node. Subsequently, the centralities of networks A (C_A) and B (C_B) are obtained from the fractions of the total centrality that remain in the nodes of A and B after the connection:

$$C_A = \frac{\sum_{i=1}^{N_A}(\mathbf{u}_{T,1})_i}{\sum_{i=1}^{N_T}(\mathbf{u}_{T,1})_i}, \tag{4.2}$$

$$C_B = 1 - C_A. \tag{4.3}$$

Suppose that a networks's goal is to accumulate as much C as possible. Regarding Eqs. (4.1) and (4.2), and taking into account that $a_1 > a_{k+1}$, with $k \geq 1$ (since eigenvalues are ranked according to their value), the final outcome of the competition depends mainly on a_1: $\mathbf{u}_{T,1} \to \mathbf{u}_{A,1}$ when $a_1 \to 0$ and therefore, $C_A \to 1$, since the elements of $\mathbf{u}_{A,1}$ are zero for all nodes belonging to network B (see Appendix for details). Otherwise, C_B will grow when a_1 grows.

But, how does a_1 depend on networks A and B, and on the connector links? Inspecting the expression of a_1 (i.e., $a_1 = [\mathbf{u}_{A,1}\mathbf{P}\mathbf{u}_{B,1}]/[\lambda_{A,1} - \lambda_{B,1}]$) we can observe that it relies on two main factors: (i) the difference between the highest eigenvalues associated to both networks, $\lambda_{A,1}$ and $\lambda_{B,1}$, and (ii) $\mathbf{u}_{A,1}\mathbf{P}\mathbf{u}_{B,1}$, a quantity that is proportional to the centralities of the connector nodes when the networks are still disconnected, and to the number of connector links. Importantly, these two factors will control the distribution of centrality between the two competing networks. While (i) is independent of the connection strategy, (ii) depends crucially on the nodes that are chosen to establish connections between A and B.

This way, when the connector nodes are the most central (i.e., $\mathbf{u}_{A,1}\mathbf{P}\mathbf{u}_{B,1}$ is maximum), network B (the weakest) shows its best results in centrality. On the contrary, when the connector links join peripheral nodes of both networks, the value of a_1 reaches its minimum. Consequently, when $a_1 \rightarrow 0$, most centrality distributes over network A and therefore $\mathbf{u}_{T,1} \rightarrow \mathbf{u}_{A,1}$, leading to $C_A \rightarrow 1$. Finally, the larger the number of connector nodes, the higher the term $\mathbf{u}_{A,1}\mathbf{P}\mathbf{u}_{B,1}$, leading to an increase of a_1 and, as a consequence, to a decrease of C_A, indicating that the strong network does not benefit from multiple connections.

It is remarkable that the expression of $\mathbf{u}_{T,1}$ can be approximated, up to first order, to a linear combination of $\mathbf{u}_{A,1}$ and $\mathbf{u}_{B,1}$ (terms $k > 1$ in Eq. (4.1) are less relevant and mainly affect the connector nodes). In spite of the fact that the percentage of centrality captured by both networks is altered by introducing connector links, the distribution of centrality inside each network after the connection is therefore to some extent proportional to what it was before.

In summary, these results allow developing a general set of strategies that competitors A and B (with $\lambda_{A,1} > \lambda_{B,1}$) should follow in order to obtain as much centrality as possible after the connection. Recalling that the *strong* network is the one with the largest first eigenvalue, and the *weak* network the one with the smallest, the general rules to maximize the outcome of a network that competes for centrality tell us that:

- Connecting the most central nodes of two networks optimizes the centrality of the weak network.
- Connecting the most peripheral nodes of two networks optimizes the centrality of the strong network.
- Increasing the number of links reinforces the centrality of the weak network.

From all above, we stress that the goal of each competitor is not really to overcome the adversary, but to obtain the optimum outcome measured with the eigenvector associated to the largest eigenvalue of the interconnected network. Importantly, the strategy played by each network depends on whether its largest eigenvalue is higher or lower than its competitor, i.e., strong and weak networks must play different strategies to maximize its outcome.

4.4 Applications

As we have seen, the selection of connector nodes between networks strongly influences the eigenvector $\mathbf{u}_{T,1}$ of the interconnected network and how its elements are distributed between the two networks forming it. Since the eigenvector centrality of the nodes is given by the eigenvector $\mathbf{u}_{T,1}$, the competition for $\mathbf{u}_{T,1}$ between two networks can be interpreted as a struggle for acquiring the highest possible centrality for the nodes inside a network. Interestingly, $\mathbf{u}_{T,1}$ may also contain information about the dynamical processes undergoing inside the network. In this section, we will show two particular examples in population dynamics and disease spreading. We will see how the previous strategies can be interpreted as a way of maximizing the outcome of a dynamical process and that this can be done by just looking at $\mathbf{u}_{T,1}$.

Population Dynamics

A variety of dynamical processes occurring on a network can be mathematically described as

$$\mathbf{n}(t+1) = \mathbf{M}\,\mathbf{n}(t)\,, \tag{4.4}$$

where $\mathbf{n}(t)$ is a vector whose components give the state of each node at time t (for example, the population of individuals at each node), and $\mathbf{M}$, with $M_{ij} \geq 0$, is a connectivity matrix that contains the peculiarities of the dynamical process (usually named as "transition matrix" in this context).

$\mathbf{M}$ is a primitive matrix. For this reason, its largest eigenvalue is positive, it verifies that $\lambda_1 > |\lambda_i|$, $\forall\, i > 1$, and its associated eigenvector is also positive (i.e., all its elements are positive). After t steps, the state of the system is given by

$$\mathbf{n}(t) = \mathbf{M}^t\mathbf{n}(0) = \sum_{i=1}^{m}(\mathbf{n}(0)\cdot\mathbf{u}_i)\lambda_i^t\mathbf{u}_i\,, \tag{4.5}$$

where $\mathbf{n}(0)$ is the initial condition, $\mathbf{u}_i$ the i−th eigenvectors of $\mathbf{M}$, and m the size of the network. As we consider M to be a real symmetric matrix, $\mathbf{u}_i$ for $i = 1, 2, \ldots, m$ can be conveniently chosen so as to form an orthonormal basis that permits the spectral decomposition above.

From Eq. (4.5) we obtain that the system evolves towards an asymptotic state independent of the initial condition and proportional to the first eigenvector $\mathbf{u}_1$,

$$\lim_{t\to\infty}\left(\frac{\mathbf{n}(t)}{(\mathbf{n}(0)\cdot\mathbf{u}_1)\lambda_1^t}\right) = \mathbf{u}_1\,, \tag{4.6}$$

while its associated eigenvalue λ_1 yields the growth rate at the asymptotic equilibrium. If $\mathbf{n}(t)$ is normalized such that $|\mathbf{n}(t)| = 1$ after each iteration, $\mathbf{n}(t) \to \mathbf{u}_1$ when $t \to \infty$. Therefore, there is a correspondence between the eigenvector centrality and the asymptotic state of the system at equilibrium: both quantities are proportional to the eigenvector $\mathbf{u}_1$ associated to the largest eigenvalue of the transition matrix $\mathbf{M}$.

Let us discuss one specific example showing the evolution of a population of genomes (e.g. RNA sequences) that duplicates and mutates inside a genotype network, where each node represents a different sequence. Two nodes are linked if they differ in only one nucleotide, and therefore one sequence can evolve from one node to the other via point mutations. At each node i of the network, we consider a certain population n_i. At each time step: (i) the population n_i replicates with a growing rate $R > 1$, (ii) its daughter individuals leave the node with probability μ, being $0 < \mu \leq 1$, and (iii) the parameter S controls how probable it is for an individual to remain alive after leaving a node (see Fig. 4.2a for a qualitative description of the process). The transition matrix describing the evolution of the

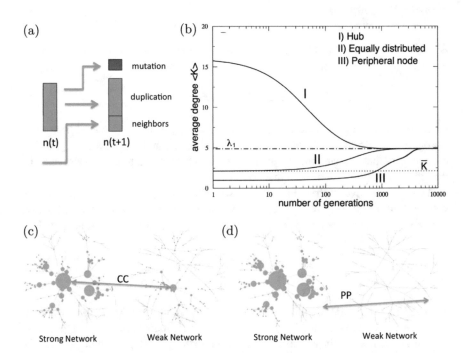

Fig. 4.2 Evolutionary dynamics of a population of genomes. (**a**) Schematic representation of the evolution of the vector state of the system $\mathbf{n}(t)$ when the population spreads on a single network of genomes. The population evolves through a duplication+mutation process, each node sending/receiving population from its neighbors when mutation occurs. (**b**) The evolution of the average degree $\langle K \rangle$ of the population shows that the final distribution is independent from the initial conditions, and higher that the average node degree $\bar{K}$. Three different initial distributions are considered: (*I*) the whole population placed at the most central node ("hub"), (*II*) uniformly distributed over the network, and (*III*) placed at the most peripheral node. (**c**) and (**d**) Evolution of the population when two networks are connected through the most central nodes (CC) and two peripheral nodes (PP) respectively. While in the CC configuration the weak network is able to retain 14.6 % of the population, in the PP case the population is almost completely absorbed by the strong network and only 10^{-4} % stays in the weak one. The parameter values are $\lambda_{A,1} = 1.9135$, $\lambda_{B,1} = 1.9109$, $R = 2$, $\mu = 0.1$, and $S = 36$ (Note that the networks used are artificial examples that verify the basic topological properties of genotype networks but do not represent real cases; see [30, 33] for more details)

system is given by [30]:

$$\mathbf{M} = (R - \mu)\mathcal{I} + \frac{\mu}{S}\mathbf{G},\tag{4.7}$$

where $\mathbf{G}$ is the adjacency matrix (i.e., $G_{ij} = 1$ if nodes i and j are connected and zero otherwise) and $\mathcal{I}$ is the identity matrix (i.e., $\mathcal{I}_{ij} = 1$ if $i = j$ and zero otherwise).

Within this framework, the eigenvector $\mathbf{u}_1$ of the matrix $\mathbf{M}$ yields the final distribution of the population at the stationary state. As expected from the reasoning above, the final distribution does not depend on the initial conditions as illustrated in Fig. 4.2b, where the population is initially distributed in three different ways. Furthermore, the average degree of the stationary population $\langle K \rangle$ is given by the largest eigenvalue λ_1, which fulfills that $\lambda_1 \geq \bar{K}$, being $\bar{K}$ the average degree (number of connections) of the nodes in the network [31].

Now, let us analyze the evolution of the population when two (sub)networks are joined through a connector link. This situation could resemble, for example, the evolution on one very modular RNA neutral network (in [32] the high modularity of such networks was recently analyzed), or two different RNA neutral networks connected via a unique link, representing each neutral network A and B the total set of sequences that fold in two different secondary structures [33]. In Fig. 4.2c, d we observe that the election of the adequate link between those two networks has critical consequences on the population accumulated at each network. Following the rules explained in the previous section, the weak network benefits from the CC connection, acquiring 14.6 % of the total population ($C_B = 0.146$). This is the best outcome that the weak network would be able to achieve when connecting through one link. On the contrary, when the PP strategy is followed, the strong network absorbs the majority of the population and the weak network remains virtually empty ($C_B = 10^{-6}$).

Spreading Processes: The SI Model

The highly developed mathematical modeling and statistical physics analysis of spreading processes have successfully described the existence of, for example, fixed points, phase transitions or spreading thresholds [34]. Among the different examples of spreading, such as rumor spreading or packet transmission through the WWW, disease spreading has been studied the most [35]. The prediction of disease evolution and the dynamics of contagions have been analyzed with a diversity of models which combine both the state of the system at different scales (from the individual to the whole population) and the structure of connections between individuals [34]. In this section our focus will be on how the structure of the network of contacts between individuals affects the probability of individuals being infected by a disease. Several works have investigated how the network topology constrains the epidemic dynamics and, more specifically, the outbreak of a disease and the properties of the epidemics in equilibrium [12, 13]. Nevertheless, less attention has been paid to the fact that social networks are typically organized in modules (or subnetworks), which interact between them through certain connector links. What is the effect of the connector links on the spreading of a disease through different subnetworks? As we are going to see, the concepts and tools defined in the previous sections will help us to understand this issue. With this aim, we are going to implement a specific disease model, the Susceptible-Infected (SI) model [36], over two networks that interact by creating interlinks as it is explained in Fig. 4.1. But first of all let us describe in detail the model and its implementation on a single network.

The SI model distinguishes two different states of the individuals: Susceptible (S) of acquiring the disease and Infected (I). When a susceptible individual (i.e., a person prone to be infected) meets an infected one, it will acquire the disease with a certain probability, which is controlled by the spreading rate β:

$$S + I \xrightarrow{\beta} 2I, \tag{4.8}$$

Next, we construct a network where the nodes are individuals and the links account for interactions between them. The connectivity matrix $\mathbf{M}$ of the network contains the connections between individuals (i.e., $M_{ij} = 1$ if two individuals are connected, and zero otherwise). The probability that a node (i.e., a person) k becomes infected is given by $I_k(t)$, while $S_k(t) = 1 - I_k(t)$ is the probability of it being susceptible (i.e., not infected). The network structure strongly influences the probability that node k becomes infected between times t and $t + dt$, as it is proportional to the number of neighbors that are already infected $\beta \sum_j M_{kj} I_j$. Since only susceptible individuals can get infected, the dynamics of $S_k(t)$ and $I_k(t)$ can be described by a set of N differential equations, N being the total number of individuals:

$$\frac{dS_k}{dt} = -\beta S_k \sum_j M_{kj} I_j = -\beta S_k \sum_j M_{kj}(1 - S_j), \tag{4.9}$$

$$\frac{dI_k}{dt} = \beta S_k \sum_j M_{kj} I_j = \beta(1 - I_k) \sum_j M_{kj} I_j, \tag{4.10}$$

with $S_k + I_k = 1$. If the disease starts from a small number of nodes, in the limit of large system size N and ignoring quadratic terms, Eq. (4.10) becomes:

$$\frac{dI_k}{dt} = \beta \sum_j M_{kj} I_j, \tag{4.11}$$

which in matrix form reads

$$\frac{d\mathbf{I}}{dt} = \beta \mathbf{M} \mathbf{I},\tag{4.12}$$

$\mathbf{I}$ being a vector of components I_k. The temporal evolution of $\mathbf{I}$ can be expressed as a linear combination of the eigenvectors $\mathbf{u}_k$ of the connectivity matrix $\mathbf{M}$:

$$\mathbf{I}(t) = \sum_{k=1}^{N} a_k(t)\mathbf{u}_k,\tag{4.13}$$

where $\mathbf{u}_k$ is the eigenvector associated with the eigenvalue λ_k of $\mathbf{M}$. Then

$$\frac{d\mathbf{I}(t)}{dt} = \sum_{k=1}^{N} \frac{da_k(t)}{dt}\mathbf{u}_k = \beta \mathbf{M} \sum_{k=1}^{N} a_k(t)\mathbf{u}_k = \beta \sum_{k=1}^{N} \lambda_k a_k(t)\mathbf{u}_k.\tag{4.14}$$

Comparing the terms that multiply $\mathbf{u}_k$, we obtain:

$$\frac{da_k}{dt} = \beta \lambda_k a_k,\tag{4.15}$$

which has the solution

$$a_k(t) = a_k(0)e^{\beta \lambda_k t}.\tag{4.16}$$

If we substitute Eq. (4.16) into Eq. (4.13) we obtain the following expression for $\mathbf{I}(t)$:

$$\mathbf{I}(t) = \sum_{k=1}^{N} a_k(0)e^{\beta \lambda_k t}\mathbf{u}_k.\tag{4.17}$$

Since the largest eigenvalue λ_1 dominates over the others, we can approximate the infected population as

$$\mathbf{I}(t) \sim e^{\beta \lambda_k t}\mathbf{u}_1.\tag{4.18}$$

Thus, for $t \to \infty$ the exponential term leads to $\mathbf{I} \to 1$, i.e. the whole population gets infected at the final state. Nevertheless, for low to intermediate time scales ($t \ll \infty$), it is $\mathbf{u}_1$, i.e. precisely the eigenvector centrality of the nodes, that controls the distribution of probabilities of getting infected.

As explained in Sect. 4.2, the properties of the eigenvector $\mathbf{u}_{T,1}$ when two networks A and B are connected depend on the kind of interlink. If we consider two networks of individuals and want to understand how the distribution of the probability of being infected depends on the kind of connection between the two

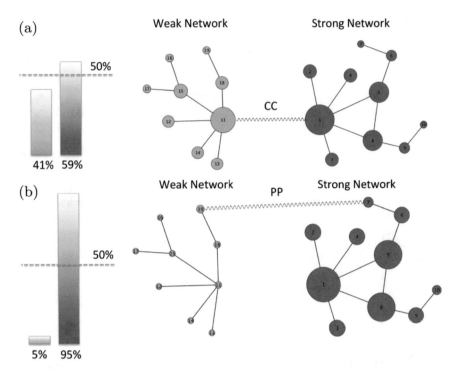

Fig. 4.3 Probability of being infected by a disease (SI model), at the beginning of the spreading process, for two interconnected networks. Two social networks based on romantic connections between young students [37] are connected through a **(a)** CC and **(b)** PP connection. Node size is proportional to the probability of being infected (obtained from $\mathbf{u}_{T,1}$) and *bars* indicate the percentage of infection risk accumulated by each network. The PP strategy leads the strong network to increase its risk of infection as compared to the weak network

networks, all strategies defined in Sect. 4.3 apply. The only difference is that, since the terms C_A and C_B are related to the probability of being infected at low to moderate times after the beginning of the epidemics, the aim of the networks will be to reduce this probability instead of increasing it. Therefore, the strategies are exactly the same as in the case of network centrality or population dynamics, but they must be applied in the opposite way.

In Fig. 4.3 we show an example of how spreading processes on interconnected networks are strongly dependent on the way the networks are linked between them. We consider two social networks based on the romantic relationships between students in an American high school [37]. Specifically, we select two subnetworks that are isolated and evaluate how a connection between one student of each subnetwork affects the probability of being infected by a disease throughout the network ensemble. With this aim, we introduce a SI spreading process with an infection rate $\beta = 0.1$ in the networks and calculate the fraction of the first eigenvector that lies within each subnetwork for a CC and a PP connection. In this particular example, the largest eigenvalues of the strong and the weak network are, respectively, $\lambda_{A,1} = 2.69$ and $\lambda_{B,1} = 2.39$. Figure 4.3a, b shows how the probability of being infected at short to moderate times is always higher in the strong network. Nevertheless, when the two most peripheral nodes of both networks are connected (Fig. 4.3b) the probability that the strong network gets infected increases dramatically. This is a situation that the strong network has to avoid, since it gets much more vulnerable to the disease than the weak network. Therefore, the PP (CC) connection is now the most harmful strategy for the strong (weak) network, while the CC (PP) connection is the safest one.

4.5 Conclusions

In this chapter we have shown that the way networks interact to form interconnected networks and, more specifically, how they choose the connector links, can have important consequences on the structural and dynamical properties of the networks [26]. A series of dynamical processes occurring on interacting networks, such as population dynamics or disease spreading, can be explained from the analysis of the spectral properties of the transition matrix, which in its turn depends on the way the networks are coupled. We have seen that it is possible to define strategies that maximize the outcome (defined in terms of the dynamical process under consideration) acquired by a certain network. As a general strategy, strong networks (i.e. those with the highest largest eigenvalue) will benefit from establishing connections between peripheral nodes. Weak networks, instead, obtain a higher benefit when the central nodes of both networks are elected as connectors.

Throughout this chapter we considered that the network's goal is to accumulate as much percentage of u_1 as possible. However, in some cases, networks may want to minimize it as in the case of disease spreading.

It is important to stress that the proposed methodology applies for processes where the final state of the system is given by the eigenvector associated to the largest eigenvalue of the transition matrix. For example, this is not the case for diffusion processes where the system dynamics is described by the (weighted) Laplacian matrix $\mathbf{L}$, obtained as $\mathbf{L} = \mathbf{W} - \mathbf{M}$, where $\mathbf{W}$ is a diagonal matrix with W_{ii} containing the total weight of node i [38].

The spectral properties of the Laplacian matrix also determine the stability of the synchronization manifold in the complete synchronization of networked systems [39]. Nevertheless, the influence of the connection strategies in the spectral properties of the Laplacian are much more difficult to interpret than in the case of the transition matrix, as there is no straightforward relation between the spectral properties of both matrices. The reader is referred to Ref. [40] for a detailed theoretical, numerical and experimental study of the effect of different connection strategies on the synchronization of an ensemble of networks.

Finally, there are other dynamical aspects that may not be explained by the analysis of either the transition or the Laplacian matrix. For example, it is expected that the complexity of the global dynamics in networks of dynamical systems that are coupled through different connection strategies will be affected by the strategy adopted in creating connections, but it is not clear at this point whether this may be related to the spectral properties of any matrix representing the coupling topology of the system. These and other problems related to networks of networks are still open and will have to be addressed in the future, showing that network interconnection is a promising subfield of network theory with potential applications in several branches of science.

Acknowledgements Authors acknowledge P.L. del Barrio for fruitful conversations and the support of MINECO (FIS2011-27569, FIS2012-38949-C03-01, FIS2013-41057-P, and FIS2014-57686). R.S.E. acknowledges UdG, Culagos (México) for financial support (PIFI 522943 (2012) and Becas Movilidad 290674-CVU-386032).

Appendix

Networks A and B, of N_A and N_B nodes and L_A and L_B links respectively, form the initially disconnected network AB of $N_A + N_B$ nodes and $I_A + L_B$ links. We connect them through L connector links to create a new interconnected network T of $N_T = N_A + N_B$ nodes and $L_T = L_A + L_B + L$ links. For convenience, the nodes of network A are numbered from $i = 1$ to N_A and the nodes of network B from $i = N_A + 1$ to $N_T = N_A + N_B$. The adjacency matrix $\mathbf{G_{AB}}$ of the disconnected network consists of two diagonal blocks corresponding to $\mathbf{G_A}$ and $\mathbf{G_B}$. The relation between the transition matrix $\mathbf{M_{AB}}$, also formed by two blocks, and $\mathbf{G_{AB}}$, depends on the peculiarities of the process. Note that the eigenvectors of $\mathbf{M_A}$ and $\mathbf{M_B}$ are related to those of $\mathbf{M_{AB}}$ as follows: Let us call $\mathbf{x}_{A,i}$ ($i = 1, \ldots, N_A$) and $\mathbf{x}_{B,j}$ ($j = 1, \ldots, N_B$) the eigenvectors associated to the eigenvalues $\lambda_{A,i}$ and $\lambda_{B,j}$ of matrices $\mathbf{M_A}$ and $\mathbf{M_B}$ respectively. Note that the N_A eigenvectors $\mathbf{x}_{A,i}$ are of length N_A, the N_B eigenvectors $\mathbf{x}_{B,j}$ are of length N_B, and the eigenvectors of $\mathbf{M_{AB}}$ are of length N_T. The first $i = 1, \ldots, N_A$ eigenvectors of $\mathbf{M_{AB}}$ verify $(\mathbf{u}_{AB,i})_k = (\mathbf{x}_{A,i})_k$ for $k \leq N_A$ and $(\mathbf{u}_{AB,i})_k = 0$ for $k > N_A$. Therefore, $\lambda_{AB,i} = \lambda_{A,i}$ for $i = 1, \ldots, N_A$. The eigenvectors $i = N_A + 1, \ldots, N_T$ of $\mathbf{M_{AB}}$ verify $(\mathbf{u}_{AB,i})_k = 0$ for $k \leq N_A$ and $(\mathbf{u}_{AB,i})_k = (\mathbf{x}_{B,i})_{k-N_A}$ for $k > N_A$. Therefore, $\lambda_{AB,i} = \lambda_{B,i-N_A}$ for $i = N_A + 1, \ldots, N_T$. For simplicity in the following calculations, due to their evident relation with the eigenvectors of $\mathbf{M_A}$, we denote eigenvectors $\mathbf{u}_{AB,i}$ for $i = 1, \ldots, N_A$ as $\mathbf{u}_{A,i}$. Analogously, we denote $\mathbf{u}_{AB,i+N_A}$ for $i = 1, \ldots, N_B$ as $\mathbf{u}_{B,i}$.

Considering the addition of interlinks as represented by the symmetric matrix $\mathbf{P}$ (with non-zero entries in the off-diagonal blocks of elements (i, j) with $i \leq N_A$ and $j > N_A$ and $i > N_A$ and $j \leq N_A$) to be a small perturbation of parameter ϵ, and Taylor-expanding the largest eigenvalue of M_T and its associated eigenvector around those of M_{AB}, we obtain

$$\mathbf{M_T u}_{T,1} = \lambda_{T,1} \mathbf{u}_{T,1} \tag{4.19}$$

where

$$M_T = M_{AB} + \epsilon P, \tag{4.20}$$

$$u_{T_1} = u_{A,1} + \epsilon v_1 + \epsilon^2 v_2 + o(\epsilon^3), \tag{4.21}$$

$$\lambda_{T,1} = \lambda_{A,1} + \epsilon t_1 + \epsilon^2 t_2 + o(\epsilon^3). \tag{4.22}$$

Taking into account that (i) $|u_{T,1}| = 1 \Rightarrow u_{A,1} \cdot v_1 = 0$ and $u_{A,1} \cdot v_2 = 0$, and (ii) $u_{A,1} P u_{A,1} = 0$ because $(u_{A,1})_i = 0$ for $i > N_A$, we include Eqs. (4.20–4.22) in Eq. (4.19), premultiply by $u_{A,1}$ and equate the terms of the same order in ϵ. Considering that point (i) above, in its turn, implies that v_1 and v_2 can be expressed as linear combinations of the other eigenvectors of M_T, which are orthogonal to $u_{A,1}$, and therefore $u_{A,1} \cdot M_{AB} v_1 = 0$ and $u_{A,1} \cdot M_{AB} v_2 = 0$, we obtain to first order in ϵ

$$u_{A,1} \cdot (M_{AB} v_1 + P u_{A,1}) = u_{A,1} \cdot (\lambda_{A,1} v_1 + t_1 u_{A,1}) \tag{4.23}$$

$$\Rightarrow t_1 = 0 \tag{4.24}$$

$$\Rightarrow (M_{AB} - \lambda_{A,1}) v_1 = -P u_{A,1}, \tag{4.25}$$

and for order ϵ^2

$$u_{A,1} \cdot (M_{AB} v_2 + P v_1) = u_{A,1} \cdot (\lambda_{A,1} v_2 + t_2 u_{A,1}) \Rightarrow t_2 = u_{A,1} P v_1. \tag{4.26}$$

The vector v_1 can be numerically obtained solving Eq. (4.25). However, it can also be analytically expressed as

$$v_1 = \sum_{k=1}^{N_T} c_k u_{AB,k} = \sum_{k=1}^{N_A} c_k u_{A,k} + \sum_{k=N_A+1}^{N_T} c_k u_{B,k-N_A}. \tag{4.27}$$

We know $c_1 = 0$ because $u_{A,1} \cdot v_1 = 0$. Including Eq. (4.27) in Eq. (4.25), and multiplying both sides by $u_{AB,k}$ from the left, we obtain $c_k = 0$ for $1 < k \le N_A$ (because $u_{A,k} P u_{A,1} = 0 \, \forall k$) and $c_k = \frac{u_{A,1} P u_{B,k-N_A}}{\lambda_{A,1} - \lambda_{B,k-N_A}}$ for $k > N_A$. All this yields

$$v_1 = \sum_{k=1}^{N_B} \frac{u_{A,1} P u_{B,k}}{\lambda_{A,1} - \lambda_{B,k}} u_{B,k}, \tag{4.28}$$

and including Eqs. (4.28) and (4.26) in Eq. (4.22), we finally obtain

$$u_{T,1} = u_{A,1} + \epsilon \sum_{k=1}^{N_B} \frac{u_{A,1} P u_{B,k}}{\lambda_{A,1} - \lambda_{B,k}} u_{B,k} + o(\epsilon^2), \tag{4.29}$$

References

1. Newman, M.E.J.: The structure and function of complex networks. SIAM Rev. **45**, 167–256 (2003)
2. Newman, M.E.J., Barabási, A.-L., Watts, D.J.: The Structure and Dynamics of Networks. Princeton University Press, Princeton/Oxford (2006)
3. Boccaletti, S., Latora, V., Moreno, Y., Chavez, M., Hwang, D.: Complex networks: structure and dynamics. Phys. Rep. **424**, 175–308 (2006)
4. Newman, M.E.J.: Networks: An Introduction. Oxford University Press, New York (2010)
5. Albert, R., Jeong, H., Bararabási, A.-L.: Error and attack tolerance of complex networks. Nature **406**, 378 (2000)
6. Callaway, D.S., Newman, M.E.J., Strogatz, S.H., Watts, D.J.: Network robustness and fragility: percolation on random graphs Phys. Rev. Lett. **85**, 5468 (2000)
7. Cohen, R., Erez, K., ben Avraham, D., Havlin, S.: Resilience of the internet to random breakdowns. Phys. Rev. Lett. **85**, 4626 (2000)
8. Cohen, R., Erez, K., ben Avraham, D., Havlin, S.: Breakdown of the internet under intentional attack. Phys. Rev. Lett. **86**, 3682 (2001)
9. Kleinberg, J.M.: Navigation in the small world. Nature **406**, 845 (2000)
10. Motter, A.E., Zhou, C.S., Kurths, J.: Enhancing complex-network synchronization. Europhy. Lett. **69**, 334 (2005)
11. Chavez, M., Hwang, D.U., Amman, A., Hentschel, H.G.E., Boccaletti, S.: Synchronization is enhanced in weighted complex networks. Phys. Rev. Lett. **94**, 218701 (2005)
12. Pastor-Satorras, R., Vespignani, A.: Epidemic spreading in scale-free networks. Phys. Rev. Lett. **86**, 3200 (2001)
13. Pastor-Satorras, R., Vespignani, A.: Epidemic dynamics and endemic states in complex networks. Phys. Rev. E **63**, 066117 (2001)
14. Zanette, D.: Critical behavior of propagation on small-world networks. Phys. Rev. E **64**, 050901 (2001)
15. Liu, Z., Lai, Y.C., Ye, N.: Propagation and immunization of infection on general networks with both homogeneous and heterogeneous components. Phys. Rev. E **67**, 031911 (2003)
16. Solé, R.V., Valverde, S.: Information transfer and phase transitions in a model of internet traffic. Phys. A **289**, 595–605 (2001)
17. Arenas, A., Díaz Guilera, A., Guimerá, R.: Communication in networks with hierarchical branching. Phys. Rev. Lett. **86**, 3196–3199 (2001)
18. Guimerá, R., Arenas, A., Díaz Guilera, A., Giralt, F.: Dynamical properties of model communication networks. Phys. Rev. E **66**, 026704 (2002)
19. Barrat, A., Barthélemy, M., Vespignani, A.: Dynamical processes on networks. Cambridge University Press, Cambridge (2008)
20. Gross, T., Blasius, B.: Adaptive networks: a review. J. R. Soc.: Interface **5**, 259–271 (2008)
21. Gao, J., Buldyrev, S., Havlin, S., Stanley, H.: Robustness of a network of networks. Phys. Rev. Lett. **107**, 1–5 (2011)
22. Quill, E.: When networks network. ScienceNews, September 22nd **182**, 6 (2012)
23. Danon, L., Duch, J., Arenas, A., Díaz-Guilera, A.: Community structure identification. In: Large Scale Structure and Dynamics of Complex Networks: From Information Technology to Finance and Natural Science, pp. 93–113. World Scientific, Singapore (2007)
24. Fortunato, S.: Community detection in graphs. Phys. Rep. **486**, 75–174 (2010)
25. Canright, G.S., Engo-Monson, K.: Spreading on networks: a topographic view. Complexus **3**, 131–146 (2006)
26. Aguirre, J., Papo, D., Buldú, J.M.: Successful strategies for competing networks. Nat. Phys. **9**, 230 (2013)
27. Marcus, R.A.: Brief comments on perturbation theory of a nonsymmetric matrix: the GF matrix. J. Phys. Chem. A **105**, 2612–2616 (2001)

28. Restrepo, J.G., Ott, E., Hunt, B.R.: Characterizing the dynamical importance of network nodes and links. Phys. Rev. Lett. **97**, 094102 (2006)
29. Chauhan, S., Girvan, M., Ott, E.: Spectral properties of networks with community structure. Phys. Rev. E **80**, 056114 (2009)
30. Aguirre, J., Buldú, J.M., Manrubia, S.C.: Evolutionary dynamics on networks of selectively neutral genotypes: effects of topology and sequence stability. Phys. Rev. E **80**, 066112 (2009)
31. van Nimwegen, E., Crutchfield, J.P., Huynen, M.: Neutral evolution of mutational robustness. Proc. Natl. Acad. Sci. USA **96**, 9716–9720 (1999)
32. Capitán, J.A., Aguirre, J., Manrubia, S.: Dynamical community structure of populations evolving on genotype networks. Chaos, Solitons & Fractals **72**, 99–106 (2014)
33. Aguirre, J., Buldú, J.M., Stich, M., Manrubia, S.C.: Topological structure of the space of phenotypes: the case of RNA neutral networks. PLoS ONE **6**(10), e26324 (2011)
34. Barrat, A., Barthélemy, M., Vespignani, A.: Dynamical processes in complex networks. Cambridge University Press, New York (2008)
35. Daley, D.J., Gani, J.: Epidemic Modeling: An Introduction. Cambridge University Press, New York (2005)
36. Brauer, F., Castillo-Chávez, C.: Mathematical Models in Population Biology and Epidemiology. Springer, New York (2001)
37. Bearman, P., Moody, J., Stovel, K.: Chains of affection: the structure of adolescent romantic and sexual networks. Am. J. Sociol. **110**, 44–99 (2004)
38. Radicchi, F., Arenas, A.: Abrupt transition in the structural formation of interconnected networks. Nat. Phys. **9**, 717–720 (2013)
39. Arenas, A., Díaz-Guilera, A., Kurths, J., Moreno, Y., Zhou, C.: Synchronization in complex networks. Phys. Rep. **469**, 93–153 (2008)
40. Aguirre, J., Sevilla-Escoboza, R., Gutiérrez, R., Papo, D., Buldú, J.M.: Synchronization of interconnected networks: the role of connector nodes. Phys. Rev. Lett. **112**, 248701 (2014)

Chapter 5
Vulnerability of Interdependent Networks and Networks of Networks

Michael M. Danziger, Louis M. Shekhtman, Amir Bashan, Yehiel Berezin, and Shlomo Havlin

Abstract Networks interact with one another in a variety of ways. Even though increased connectivity between networks would tend to make the system more robust, if dependencies exist between networks, these systems are highly vulnerable to random failure or attack. Damage in one network causes damage in another. This leads to cascading failures which amplify the original damage and can rapidly lead to complete system collapse.

Understanding the system characteristics that lead to cascading failures and support their continued propagation is an important step in developing more robust systems and mitigation strategies. Recently, a number of important results have been obtained regarding the robustness of systems composed of random, clustered and spatially embedded networks.

Here we review the recent advances on the role that connectivity and dependency links play in the robustness of networks of networks. We further discuss the dynamics of cascading failures on interdependent networks, including cascade lifetime predictions and explanations of the topological properties which drive the cascade.

5.1 Background: From Single Networks to Networks of Networks

As the ability to measure complex systems evolved, driven by enhanced digital storage and computation abilities in the 1990s, researchers discovered that network topology is important and not trivial. New structures were observed and new

M.M. Danziger (✉) • L.M. Shekhtman • Y. Berezin • S. Havlin
Department of Physics, Bar Ilan University, Ramat Gan, Israel
e-mail: michael.danziger@biu.ac.il; lsheks@gmail.com; bereziny@gmail.com;
havlin@ophir.ph.biu.ac.il

A. Bashan
Channing Division of Network Medicine, Brigham Women's Hospital and Harvard Medical School, Boston, MA, USA
e-mail: amir.bashan@channing.harvard.edu

© Springer International Publishing Switzerland 2016   79
A. Garas (ed.), *Interconnected Networks*, Understanding Complex Systems,
DOI 10.1007/978-3-319-23947-7_5

models proposed to explain them. Scale-free networks dominated by hubs [1, 2], small-world networks which captured the familiar "six degrees of separation" idea [3, 4], ideas of communities and clustering, and countless other variations [5, 6] were discovered and analyzed. Network topologies were shown to be very different from the abstractions of classical graph theory [7–9] in many real systems and yet important calculations, predictions and measurements could still be executed. Looking to the topology of networks provided new insights into epidemiology [10], marketing [11], percolation [12], traffic [13], and climate studies [14, 15] amongst many others.

One of the most important properties of a network that was studied was its vulnerability to the failure of a subset of its nodes. Utilizing percolation theory, network robustness can be studied via the fraction of nodes in its largest connected component P_∞ which is taken as a proxy for functionality of the network [16, 17]. Consider, for example, a telephone network composed of telephone lines and retransmitting stations. If $P_\infty \sim 1$ (the entire system), then there is a high level of connectivity in the system and information from one part of the network is likely to reach any other part. If, however, $P_\infty \sim 0$, then information in one part cannot travel far and the network must be considered nonfunctional. Even if $P_\infty \sim 1$, some nodes may be detached from the largest connected component and those nodes are considered nonfunctional. We use the term *giant connected component* (GCC) to refer to P_∞ when it is of order 1. Percolation theory is concerned with determining $P_\infty(p)$ after a random (or targeted) fraction $1 - p$ of nodes (or edges) are disabled in the network. Typically, $P_\infty(p)$ undergoes a second-order transition at a certain value p_c: for $p > p_c$, $P_\infty(p) > 0$ and it approaches zero as $p \rightarrow p_c$ but for $p < p_c$, $P_\infty(p) \equiv 0$. Thus there is a discontinuity in the derivative $P'_\infty(p)$ at p_c even though the function itself is continuous. It is in this sense that the phase transition is described as second-order [18]. It was shown, for example, that scale-free networks (SF)–which are extremely ubiquitous in nature–have $p_c = 0$ as long as the degree distribution has a sufficiently long tail [12]. This is in marked contrast to Erdős-Rényi (ER) networks ($p_c = 1/\langle k \rangle$) and 2D square lattices ($p_c \approx 0.5927$ [17]) and helps to explain the surprising robustness of many systems (e.g. the internet) with respect to random failures [12, 19].

However, in reality, networks rarely appear in isolation. In epidemiology, diseases can spread within populations but can also transition to other populations, even to different species. In transportation networks, there are typically highway, bus, train and airplane networks covering the same areas but behaving differently [20]. Furthermore, the way in which one network affects another is not trivial and often specific nodes in one network interact with specific nodes in another network. This leads to the concept of interacting networks in which links exist between nodes within a single network as well as across networks. Just as ideal gases–which by definition are comprised of non-interacting particles–lack emergent critical phenomena such as phase transitions, we will see that the behavior of interacting networks has profound emergent properties which do not exist in single networks.

Since networks interact with one another selectively (and not generally all networks affecting all other networks), we can describe *networks of networks* (NoN) with topologies between networks that are similar to the topology of nodes in a single network.

Multiplex networks are interconnected networks in which the identity of the nodes is the same across different networks but the links are different [21–23]. Multiplex networks were first introduced to describe a person who participates in multiple social networks [24]. For instance, the networks of phone communication and email communication between individuals will have different topologies and different dynamics though the actors will be the same [25]. Also, each online social network shares the same individuals though the network topologies will be very different depending on the community which the social network represents.

When discussing networks of networks, a natural question is: why describe this phenomenon as a "interconnected networks?" If we are dealing with a set of nodes and links then no matter how it is partitioned it is still a network. Each description of interacting networks will answer this question differently but any attempt to describe a network of networks will be predicated on a claim that more is different—that by splitting the overall system into component networks, new phenomena can be uncovered and predicted. One way of describing the interaction between networks which yields qualitatively new phenomena is *interdependence*. This concept has been studied in the context of critical infrastructure and been formalized in several engineering models [26, 27] (see Fig. 5.1). However, as a theoretical property of interacting networks, interdependence was first introduced in a seminal study by Buldyrev et al. in 2010 [28]. This review will focus on the theoretical framework and wealth of new phenomena discovered in interdependent networks. Some parts of this review first appeared in the proceedings of NDES 2014 [29].

5.2 Interdependence: Connectivity and Dependency Links

The fundamental property which characterizes interdependent networks is the existence of two qualitatively different kinds of links: *connectivity* links and *dependency* links [28, 30–32] (see Fig. 5.1). The connectivity links are the links which we are familiar with from single network theory and they connect nodes within the same network. They typically represent the ability of some quantity (information, electricity, traffic, disease etc.) to flow from one node to another. From the perspective of percolation theory, if a node has multiple connectivity links leading to the GCC, it will only fail if all of those links cease to function. Dependency links, on the other hand, represent the idea that for a node to function, it requires support from another node which, in general, is in another network. In such a case, if the supporting node fails, the dependent node will also fail–even if it is still connected to the GCC in its network. If one network *depends on* and *supports* another network, we describe that pair of networks as interdependent. Interdependence is a common feature of critical infrastructure (see Fig. 5.1) and

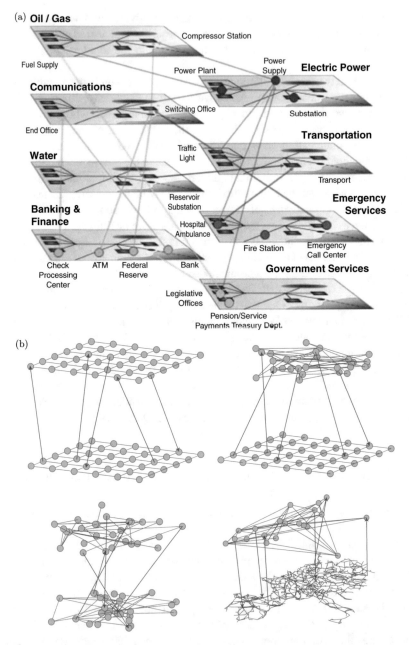

Fig. 5.1 An example of interdependent critical infrastructure systems and several modelled interdependent networks. (**a**) Schematic representation of interdependent critical infrastructure networks after [33]. (**b**) Illustration of interdependent networks composed of connectivity links (in *blue*, within the networks) and dependency links (in *red*, between the networks). Clockwise from upper-left: coupled lattices, a lattice coupled with a random regular (RR) network, two coupled RR networks and an RR network coupled to a real-world power grid (After [34])

many multiplex networks. Often whatever causes a node to stop functioning in one layer will also disable it in other layers. Indeed, the percolation properties of interdependent networks describe the typical behavior in multiplex networks as well [28]. The properties of interdependence can affect a network's function in a variety of ways but here we focus on the response of a network of interdependent networks to the failure of a subset of its nodes using the tools of percolation theory [6]. We refer the reader to recent general reviews for other descriptions of interacting networks [24, 35, 36].

Percolation on a single network is an instantaneous process but on a system of interdependent networks, the removal of a random fraction $1 - p$ of the nodes initiates a cascading failure in the following sense. Consider percolation on two interdependent networks A and B for which every node in A depends on exactly one node in B and vice versa. If we remove a fraction $1 - p$ of the nodes in A, other nodes in A which were connected to the GCC via the removed nodes will also be disabled, leaving a new GCC of size $P_\infty(p) < p$. Since all of the nodes in B depend on nodes in A, a fraction $1 - P_\infty(p)$ of the nodes in B will now be disabled via their dependency links. This will lead, in turn, to more nodes being cut off from the GCC in B and the new GCC in B will be smaller yet. This will lead to more damage in A due to the dependency links from B to A. This process of percolation and dependency damage accumulating iteratively continues until no more nodes are removed from iteration to iteration. This cascading failure is similar to the cascades described in flow and overload models on networks and the cascading failures in power grids which are linked to blackouts [37, 38]. The cascade triggered by a single node removal has been called an "avalanche"[39] and the critical properties of this process have been studied extensively [39, 40].

5.3 Interdependent Random Networks

This cascading failure was shown to lead to abrupt first-order transitions in systems of interdependent ER and SF networks that are qualitatively very different from the transitions in single networks (see Fig. 5.2). Furthermore, p_c of a pair of ER networks was shown to increase from $1/\langle k \rangle$ to $2.4554/\langle k \rangle$. Surprisingly, it was found that scale-free networks, which are extremely robust to random failure on their own [12, 19], become more vulnerable than equivalent ER networks when they are fully interdependent and for any $\lambda > 2$, $p_c > 0$. In general, a broader degree distribution leads to a higher p_c [28]. This is because the hubs in one network, which are the source of the stability of single scale-free networks, can be dependent on low degree nodes in the other network and are thus vulnerable to random damage via dependency links. These results were first demonstrated using the generating function formalism [28, 41], though it has recently been shown that the same results can be obtained using the cavity method [42].

After the first results on interdependent networks were published in 2010 [28], the basic model described above was expanded to cover more diverse systems. One

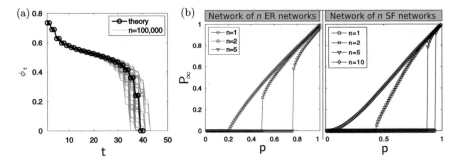

Fig. 5.2 Percolation of a network of interdependent random networks. (**a**) The fraction of viable nodes at time t for a NoN composed of 5 ER networks. The *gray lines* represent individual realizations and the *black line* is calculated analytically. After [51]. (**b**) Percolation in a NoN of ER and SF networks. Shown here is the effect of increasing the number of networks n for tree-like NoNs composed of ER and SF networks (After [51])

striking early result was that if less than an analytically calculable critical fraction q_c of the nodes in a system of two interdependent ER networks are interdependent, the phase transition reverts to the familiar second-order transition [30]. However, for scale-free networks, reducing the fraction of interdependent nodes leads to a hybrid transition, where a discontinuity in P_∞ is followed by a continuous decline to zero, as p decreases [43]. A similar transition was found when connectivity links between networks (which were first introduced in [44]) are combined with dependency links [45]. It has also been shown that the same cascading failures emerge from systems with connectivity and dependency links within a single network [46–48].

The assumption that each node can depend on only one node was relaxed in [49] and it was shown that even if a node has many redundant dependency links, the first-order transition described above can still take place. If dependency links are assigned randomly, a situation can arise in which a chain of dependency links can be arbitrarily long and thus a single failure can propagate through the entire system. To avoid this scenario, most models for interdependent networks assume uniqueness or "no feedback" which limits the length of chains of dependency links [50, 51]. For a pair of fully interdependent networks, this reduces to the requirement that every dependency link is bidirectional. Under partial dependency, this assumption is not necessary and the differences between systems with and without feedback have also been studied [34, 51].

Though both the connectivity and dependency links were treated as random and uncorrelated in Refs. [28, 30, 31, 50–52], the theory of interdependent networks has been expanded to more realistic cases. Assortativity of connectivity links was shown to decrease overall robustness [53]. Assortativity of dependency links was treated numerically [54], analytically for the case of full degree-degree correlation [55] and analytically for the general case of degree-degree correlations with connectivity or dependency links using the cavity method [42]. Interestingly, if a fraction α of the highest degree nodes are made interdependent in each network, a three-phase system with a tricritical point emerges in the α–p plane [56]. If the system is a multiplex network, there may be overlapping links, i.e., two nodes which are linked in one layer may have a tendency to be linked in other layers [57–59]. In interdependent networks this phenomenon is referred to as intersimilarity [54, 60]. Clustering, which has a negligible effect on the robustness of single networks [61], was shown to substantially reduce the robustness of interdependent networks [43, 62] and networks of networks [63].

In the following subsections, we highlight some significant methodological approaches and results from the study of cascading failures in two interdependent networks (Sect. 5.3) and from the study of networks of n interdependent networks (Sect. 5.3).

Cascading Failures in Coupled Networks

Consider two networks A and B which are partially dependent in the sense that only a fraction q_A (q_B) of the nodes in A (B) are dependent, the rest being autonomous [30]. If a fraction $1 - p$ of the nodes in A are removed, we define ψ_t' (ϕ_t') as the fraction of viable nodes at time t in network A (B). Of those nodes, the fraction which are part of the GCC is given by $\phi_t = \phi_t' g_A(\phi_t')$ ($\psi_t = \psi_t' g_B(\psi_t')$). By tracing the value of ψ_t' and ϕ_t', we can measure and predict the dynamics of the cascading failure in a system of interdependent networks. The function $g_i(p)$ can be determined analytically for ER, SF and indeed for a random network with an arbitrary degree distribution using generating functions. Using this, we can predict the size of the giant component in both networks at every time t:

$$
\begin{aligned}
\psi_1' &\equiv p \\
\phi_1' &= 1 - q_B[1 - pg_A(\psi_1')] \\
\psi_t' &= p(1 - q_A[1 - g_B(\phi_{t-1}')]) \\
\phi_t' &= 1 - q_B[1 - pg_A(\psi_{t-1}')]
\end{aligned}
\tag{5.1}
$$

Since the steady state is defined as the configuration for which $\psi_t = \psi_{t-1}$ and $\phi_t = \phi_{t-1}$ we obtain a system of two equations and two unknowns:

$$
\begin{aligned}
\psi_\infty' &= p(1 - q_A[1 - g_B(\phi_\infty')]) \\
\phi_\infty' &= 1 - q_B[1 - pg_A(\psi_\infty')]
\end{aligned}
\tag{5.2}
$$

Depending on q_i and g_i, the size of the GCC in each network will either approach zero as $p \to p_c$ in which case there will be a second-order transition or will abruptly jump to zero and there will be a first-order transition. Results of these calculations are shown in Fig. 5.2. The cascade "plateau" emerges from the analytic predictions as well as simulations. We discuss this phenomenon in greater detail in Sect. 5.4. Letting $q_A = q_B = 1$ and $g_A = g_B = g$ we recover the results from [28].

Results for Networks of Interdependent Networks

In a series of articles, Gao et al. extended the theory of pairs of interdependent networks to networks of interdependent networks with general topologies [31, 50–52]. Within this framework, analytic solutions for a number of key percolation quantities were presented including size of the GCC at each time-step t (see Fig. 5.2), the size of the GCC at steady state (see Fig. 5.2), p_c and other values.

The NoN topologies which were solved analytically include: a tree-like NoN of ER, SF or random regular (RR) networks ($q = 1$), a loop-like NoN of ER, SF or RR networks ($q \leq 1$), a star-like NoN of ER networks ($q \leq 1$) and a RR NoN of ER, SF or RR networks ($q \leq 1$). For tree-like NoNs, it was found [31, 52] that the number of networks in the NoN (n) affects the overall robustness but the specific topology of the NoN does not. In contrast, for a RR NoN the number of networks n does not affect the robustness but the degree of each network within the NoN (m) does [50, 51]. Because the topology of the loop-like and RR NoNs allows for chains of dependency links going throughout the system, there exists a quantity q_{max} above which the system will collapse with the removal of a single node, even if each network is highly connected ($p = 1$).

In a NoN, each node is a network and pairs of networks are considered linked if dependency links exist between then. We define a "NoN adjacency matrix" Q with elements q_{ij} defined as the fraction of nodes in network i that depend on nodes in network j. Recently, it was shown that the formalism developed for analytically solvable networks can be applied to NoNs for which the percolation profile of the individual networks is known only numerically [64].

For a tree-like NoN formed of n ER networks [50, 52] the size of the GCC is obtained from the self-consistent solution to

$$P_\infty = p[1 - e^{-\langle k \rangle P_\infty}]^n \tag{5.3}$$

For $n = 1$ this is the familiar second-order transition for a single ER network [7–9] but for $n = 2$ (as in [28]) or greater, there is a discontinuity in P_∞ and the transition is first-order as shown in Fig. 5.2. For fully interdependent, tree-like NoNs, the robustness decreases as n increases but is not impacted by the specific topology of the NoN. For case of partially interdependent ER networks, the specific topology does influence the robustness as shown analytically for the special case of a star-like NoN in [31]. Similar results have been obtained for trees of RR networks [50, 52] and SF networks [51].

For a loop-like NoN of partially interdependent ER networks [31, 50] the number of networks also does not affect the robustness and the GCC can be calculated as

$$P_\infty = p(1 - e^{-\langle k \rangle P_\infty})(qP_\infty - q + 1), \qquad (5.4)$$

which recovers the familiar result for single networks if $q = 0$.

A more thorough analysis of the influence of loops in NoNs appears in a random-regular network of ER networks (RR NoN of ERs). Such a system can exhibit first or second order phase transitions depending on the value of q [51]. For $q < q_c$, the transition is second-order and takes place at $p = p_c^{II}$. For $q_c < q < q_{max}$, the transition is first-order and takes place at $p = p_c^{I}$. Above q_{max}, the feedback loops enabled by the NoN topology lead to spontaneous collapse, even in a fully connected network. The mutual GCC for an RR NoN of ERs is

$$P_\infty = \frac{p}{2^m} \left(1 - e^{-kP_\infty}\right) \left(1 - q + \sqrt{(1-q)^2 + 4qP_\infty}\right)^m \qquad (5.5)$$

from which we can derive

$$p_c^{II} = \frac{1}{\langle k \rangle (1 - q)^m} \qquad (5.6)$$

and

$$q_c = \frac{k + m - \sqrt{m^2 + 2km}}{k}. \qquad (5.7)$$

The values of q_{max} and p_c^{I} can also be derived analytically from Eq. (5.5), but require several intermediate results. We refer the interested reader to the original derivation in Gao et al. [51].

In light of these results, we can now see that single network percolation is simply a limiting case of NoN percolation theory. These results have been recently reviewed in [32] and [65].

5.4 Critical Dynamics and the Cascade "Plateau"

Because the phase transition in interdependent networks is characterized by a cascading process, it has a duration which is determined by the number of iterations and is referred to in the literature as *NOI* or τ. In random networks at criticality, the size of the GCC decreases from iteration to iteration via an initial quick drop in size, followed by a long period of very little change (the "plateau") and finally a fast collapse, see Fig. 5.2. A similar plateau with different scaling behavior appears in spatially embedded networks with random dependency links, as discussed in [66].

Explaining the critical properties of this plateau is an important result of recent research [40].

Zhou et al. [40] showed that at the critical point of a given realization, p_c, the duration of the plateau, and thus the cascade (τ) scales with system size as

$$\tau \sim N^{1/3} \tag{5.8}$$

in ER networks and thus diverges in the thermodynamic limit. The value of the scaling exponent can be derived directly from universality arguments alone. If we consider that ER networks are describable with mean-field theory, they are thus in the same universality class as 6-dimensional lattice percolation (i.e., $N = L^6 \leftrightarrow L = N^{1/6}$). If we consider the branching process which characterizes the bulk of the cascading failure (Fig. 5.3), we can see that its duration will scale linearly with the number of steps (l) required for a random walker to traverse an ER network at criticality (i.e. $\tau \sim l$). Absent interactions as is the case for mean field theory, this scales with linear system size as

$$l \sim L^2 \rightarrow \tau \sim N^{1/3}. \tag{5.9}$$

Combining Eq. (5.9) with the fact that the upper critical dimension for percolation is 6, we recover Eq. (5.8). A different derivation of this result was previously published by Zhou et al. [40]. Buldyrev et al. [28] showed that for the *mean* p_c, $\tau \sim N^{1/4}$. Zhou et al. [40] developed a theory to explain the relation of this result to Eq. (5.8) that was found for single realizations.

Recent work by Zhou et al. [40] has shed new light on the dynamics of the plateau formation at criticality (Figs. 5.2 and 5.3). When a node a_i in network A fails, it will typically cause a node to fail in network B. This may lead to further damage in B due to percolation and that damage will cause the failure of a (possibly empty) set $a' \subset A$ due to the dependency links. Thus at each iteration, there is a branching process of induced damage in each network.

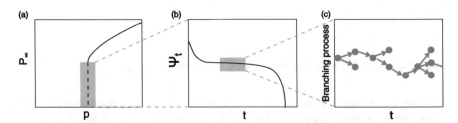

Fig. 5.3 (**a**) The first order transition in a system of interdependent networks is characterized by an abrupt drop in the size of the GCC. (**b**) On closer inspection, this jump is the product of a cascade of failures, the bulk of which is dominated by a "plateau" during which the GCC changes very little. (**c**) The plateau can be analyzed in terms of a branching process where the branching factor at time t, η_t, describes the number of failures at time t relative to the step before. During the plateau, $\eta_t \approx 1$ due to a balance of competing processes, as described in the text (After [40])

The researchers in [40] examined s_t, the number of nodes which failed at time t from the root node. They then defined the branching factor, η_t, as s_{t+1}/s_t. It was shown that η_t goes through three phases. When t is small, $\eta_t < 1$ and the branching process is decaying. This is due to the fact that the dependency damage which A carries to B causes less percolative damage in B and thus less dependency damage back in A. If the network was not also becoming more dilute in the process, then the branching process would decay and stop quickly. Indeed, for $p > p_c$ this is what happens. However, when $p = p_c$, the process continues for an infinite amount of time (in the thermodynamic limit). This is because although $\eta_t < 1$, the network becomes weaker each time nodes are removed and at criticality these processes are exactly balanced. Thus the plateau stage is a second-order percolation transition caused spontaneously by a perfect matching between the dilution of the network (which would tend to amplify the damage) and the decreasing damage due to percolation (which would decrease the damage). During this stage the branching factor is $\eta_t \approx 1$. However, due to the finite size of the system, the network eventually becomes sufficiently dilute for $\eta_t > 1$. At this point, η_t grows exponentially and the entire system collapses within a few steps.

5.5 Spatially Embedded Interdependent Networks

One of the most compelling motivations for developing a theory of interdependent networks is that many critical infrastructure networks depend on one another to function [26, 27]. Essentially all critical infrastructure networks depend on electricity to function, which is why threats like electromagnetic pulses are taken so seriously (see Fig. 5.1, Ref. [33]). The power grid itself, though, requires synchronization and control which it can only receive when the communication network is operational. One of the largest blackouts in recent history, the 2003 Italy blackout, was determined to have been caused by a cascading failure between electrical and communications networks [67].

In contrast to abstract networks, all infrastructure networks are embedded in space [20]. The nodes (e.g., power stations, communication lines, retransmitters etc.) occupy specific positions in a 2D plane and the fact that the cost of links increases with their length leads to a topology that is markedly different from random networks [68]. Thus infrastructure networks will tend to be approximately planar and the distribution of geographic link distances will be exponential with a characteristic length [69]. From universality principles, all such networks are expected to have the same general percolation behavior as standard 2D lattices [16, 69]. As such, the first descriptions of spatially embedded interdependent networks were modelled with square lattices [34, 64, 70–72] and the results have been verified on synthetic and real-world power grids [34, 71].

Analytic descriptions of percolation phenomena require the network to be "locally tree-like" and in the limit of large systems, this assumption is very accurate for random networks of arbitrary degree distribution [41]. However, lattices and other spatially embedded networks are not even remotely tree-like and analytic results on percolation properties are almost impossible to obtain [16, 17]. Therefore most of the results on spatially embedded networks are based on numerical simulations.

One of the few major analytic results for spatially embedded systems is that for interdependent lattices, if there is no restriction on the length of the dependency links then any fraction of dependency leads to a first-order transition ($q_c = 0$). In [34], it was shown that the critical fraction q_c for which the system transitions from the first-order regime to the second order regime must fulfill:

$$1 = p_c^\star q_c P'_\infty(p_c) \qquad (5.10)$$

in which $p_c^\star$ is the percolation threshold in the system of interdependent lattices, p_c is the percolation threshold in a single lattice and $P'_\infty(p)$ is the derivative of $P_\infty(p)$ for a single lattice. Since as $p \to p_c$, $P_\infty(p) = A(x - p_c)^\beta$ and for 2D lattices $\beta = 5/36$ [73], $P'_\infty(p)$ diverges as $p \to p_c$ and the only way to fulfill Eq. (5.10) is if $q_c = 0$. From universality arguments, all spatially embedded networks in $d < 6$ have $\beta < 1$ [16, 17, 69] and thus all systems composed of interdependent spatially embedded networks (in $d < 6$) with random dependency links will have $q_c = 0$. In Fig. 5.1, all of the configurations shown except the RR-RR system have $q_c = 0$.

If the dependency links are of limited length, the percolation behavior is surprisingly complex and a new spreading failure emerges. Li et al. [70] introduced the parameter r, called the "dependency length," to describe the fact that in most systems of interest the dependency links, too, will likely be costly to create and, like the connectivity links, will tend to be shorter than a certain characteristic length. In this model, dependency links between networks are selected at random but are always of length less than r (in lattice units). If $r = 0$, the system of interdependent lattices behaves identically to a single lattice. If $r = \infty$, the dependency links are unconstrained and purely random as in [34]. Li et al. [70] found that p_c as a function of r shows a sharp maximum at $r_c = 8$ which is explained by the correlation length of percolation. Moreover, as long as r is below a critical length r_c, the transition is second-order but for $r > r_c$ the transition is first order (See Fig. 5.4). The first-order transition for spatially embedded interdependent networks is unique in that it is characterized by a spreading process. Once damage of a certain size emerges at a given place on the lattice, it will begin to propagate outwards and destroy the entire system (See Fig. 5.4).

If the dependency is reduced from $q = 1$ to lower values, it is found that r_c increases and diverges at $q = 0$, consistent with the result from [34] that $q_c = 0$ for $r = \infty$ [72] (See Fig. 5.4).

Recently, the framework developed in [31, 50–52] was extended to general networks of spatially embedded networks in [64]. There they developed a theory for a network of spatially embedded interdependent networks with $r = \infty$ and presented simulation results for the case of finite r.

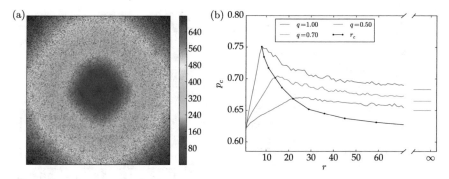

Fig. 5.4 Percolation of spatially embedded networks. (**a**) A snapshot of one lattice in a pair of interdependent lattices with nodes colored according to the time-step in which the node failed. The regularity of the color-change reflects the constant speed of the spreading failure in space (Generated for $q = 1, r = 11, L = 2900$). (**b**) The effect of r and q on p_c. As r increases, p_c increases until r reaches r_c. At that point the transition becomes first-order and p_c starts decreasing until it reaches its asymptotic value at $r = \infty$ (Both after [72])

Though the generating function for a square lattice is not analytically solvable, we do know how P_∞ behaves as a function of p for a single lattice. In [34, 70], that information was utilized to derive the theoretical mutual giant connected component for a system of two interdependent lattices. Shekhtman et al. [64] extended that theory to the case of a network composed of n lattices. Specifically, three main cases were solved: a treelike fully dependent network of lattices, a starlike partially dependent network of lattices, and a random-regular partially dependent network of lattices. Similar to the case of networks of random networks, the robustness of fully dependent tree-like spatially embedded NoNs are affected by n but not by the topology of the tree while RR NoNs are affected by m (the number of networks that each network depends on) but not by n [64]. Furthermore, the theory derived in [64] can be used to find the mutually giant connected component of any system of interdependent networks where we know the percolation profile of the individual networks.

For the case of random-regular networks there exists a certain fraction of interdependence, q_{max}, for which removing even a single node, i.e. $p \to 1$, causes the entire system to collapse [51] (see Sect. 5.3). In networks of lattices, this fraction decreases rapidly and for $m \geq 15$ only 10 % of nodes need to be interdependent for the entire system to collapse after a single node is removed [64].

The extension of analytical results from random networks to spatially embedded networks is possible only for the case in which the dependency links are purely random ($r = \infty$). As mentioned above, two fully interdependent lattices undergo a first-order transition only when $r > r_c \approx 8$ [66, 70]. This requires nodes to be dependent on their eighth nearest neighbors, which may be unlikely for a real system. Shekhtman et al. [64] showed that r_c decreases significantly as n increases (for treelike networks) and as m increases (for random-regular networks) (Fig. 5.5). They further observed that for $m \geq 15$, q_{max} is approximately

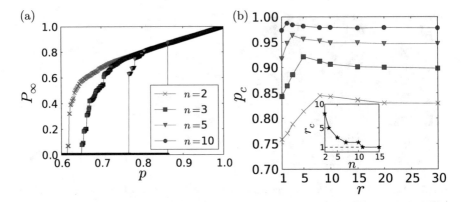

Fig. 5.5 Percolation of interdependent spatially embedded networks. (**a**) Here we observe that for fully dependent treelike NoNs with $r = 2$ the transition becomes first order as the number of networks increases. (**b**) The transition becomes first order where the p_c-r curve reaches a max. This occurs for smaller value of r as n increases. In the inset we show how r_c decreases as n increases (After [64])

independent of r. In this case, even systems with short dependency links (low r) and small fractions of dependent nodes q can collapse when only a single node is removed.

The model of spatially embedded interdependent networks was extended to the case where the ability to provide support to a node in another network requires dynamic functionality in the form of the flow of current and not just connectivity to the giant component. Process-based dependency leads to more vulnerable systems than structural dependency as described in other models. Also, the current-based model suggests that the ideas of interdependent networks can be utilized for new kinds of sensors [74].

Recently, spatially embedded interdependent networks have been modeled as multiplex networks with connectivity links of characteristic geographic length [75]. In this model, the connectivity links in each layer have lengths which are distributed exponentially. Instead of nodes in one network depending on nodes in another, each node has links in multiple networks and requires connectivity in each layer to function. This model exhibits first or second-order transitions, depending on the characteristic length of the connectivity links.

5.6   Attack and Defense of Interdependent Networks

Due to their startling vulnerabilities with respect to random failures, it is of particular interest to understand how non-random attacks affect interdependent networks and how to improve the robustness of interdependent networks through topological changes. Huang et al. [76] studied tunable degree-targeted attacks on interdependent networks. They found that even attacks which only affected low-degree nodes caused severe damage because high-degree nodes in one network can depend on low-degree nodes in another network. This framework was later expanded to general networks of networks [77].

Since high degree nodes in one network which depend on low degree nodes in another network can lead to extreme vulnerability, there have been several attempts to mitigate this vulnerability by making small modifications to the inter-network topology. Schneider et al. [78] demonstrated that selecting autonomous nodes by degree or betweenness can greatly reduce the chances of a catastrophic cascading failure. Valdez et al. have also obtained promising results by selecting a small fraction of high-degree nodes and making them autonomous [79]. These mitigation strategies are methodologically related to the intersimilarity/overlap studies discussed above [54, 60].

The theory of stochastic block models [80, 81] has been generalized to model interdependent networks and networks of networks. Using this framework, it was found that the optimal topological configuration which balances construction cost and robustness to random failure for random and interdependent networks is a core-periphery topology [82].

As we have seen, cascading failures are dynamic processes and the overall cascade lifetime can indeed be very long [40, 66]. The slowness of the process opens the door for "healing" methods allowing the dynamic recovery of failed nodes in the midst of the cascade. Recently, a possible healing mechanism along these lines has been proposed and analyzed [83].

When considering infrastructure or other spatially embedded networks, not only is the network embedded in space but failures are also expected to be geographically localized. For instance, natural disasters can disable nodes across all networks in a given area while EMP or biological attacks can disable the power grid or social network only in a given area. Geographically localized attacks of this sort have received attention in the context of single network percolation on specific networks [84] and flow-based cascading failures [85]. However, the existence of dependency between networks leads to surprising new effects. Recently, Berezin et al. [71] have shown that spatially embedded networks with dependencies can exist in three phases: stable, unstable and metastable (See Fig. 5.6). In the metastable phase, the system is robust with respect to random attacks–even if finite fractions of the system are removed. However, if all of the nodes within a critical radius r_h^c fail, it causes a cascading failure which spreads throughout the system and destroys it (See Fig. 5.6). Significantly, the value of r_h^c does not scale with system size and thus, in the limit of large systems, it constitutes a zero-fraction of the total system. A method of localized attacks on random networks was also recently studied in Shao et al. [86].

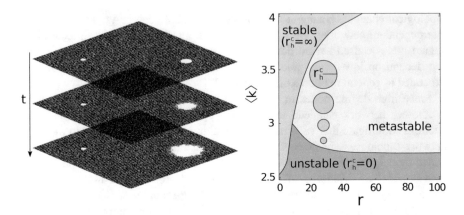

Fig. 5.6 Geographically localized attacks on interdependent networks. (**a**) The hole on the left is below r_h^c and stays in place while the hole on the right is larger than r_h^c and propagates through the system. (**b**) The phase space of localized attacks on interdependent networks. The increasing *gray circles* represent the dependence of r_h^c on $\langle k \rangle$ (Both after [71])

5.7   Applications of Networks of Networks

Many of the fields for which networks were seen as relevant models have been re-evaluated in light of the realization that interacting networks behave differently than single networks. Epidemics on interdependent and interconnected networks have received considerable attention [25, 87–90]. Economic networks composed of individuals, firms and banks all interact with one another and are susceptible to large scale cascading failures [91–93]. Interacting networks have also been found in physiological systems [94], ecology [95] and climate studies [96]. Recently, Reis et al. published an important step connecting interacting networks with fMRI measurements of brain activity [97]. Multilevel transportation networks have also been studied from the perspective of interacting networks [98]. Recently a framework for optimal recovery of interdependent networks was developed by Majdandzic et al. [99].

The breadth of applications of networks of networks is too great to address here and we refer the reader to recent reviews for more thorough treatment of applications [24, 35].

Acknowledgements We acknowledge the LINC (No. 289447) and MULTIPLEX (No. 317532) EU projects, the Deutsche Forschungsgemeinschaft (DFG), the Israel Science Foundation, ONR and DTRA for financial support.

References

1. Barabási, A.L., Albert, R.: Emergence of scaling in random networks. Science **286**(5439), 509–512 (1999)
2. Caldarelli, G.: Scale-Free Networks: Complex Webs in Nature and Technology. Oxford University Press, Oxford (2007)
3. Watts, D.J., Strogatz, S.H.: Collective dynamics of 'small-world' networks. Nature **393**(6684), 440–442 (1998)
4. Amaral, L.A.N., Scala, A., Barthélemy, M., Stanley, H.E.: Classes of small-world networks. Proc. Natl. Acad. Sci. **97**(21), 11149–11152 (2000)
5. Newman, M.: Networks: An Introduction. Oxford University Press, Oxford (2010)
6. Cohen, R., Havlin, S.: Complex Networks: Structure, Robustness and Function. Cambridge University Press, New York (2010)
7. Erdős, P., Rényi, A.: On random graphs i. Publ. Math. Debrecen **6**, 290 (1959)
8. Erdős, P., Rényi, A.: On the strength of connectedness of a random graph. Acta Mathematica Academiae Scientiarum Hungaricae **12**(1–2), 261–267 (1964)
9. Bollobás, B.: Modern Graph Theory. Graduate Texts in Mathematics. Springer, New York (1998)
10. Pastor-Satorras, R., Vespignani, A.: Epidemic spreading in scale-free networks. Phys. Rev. Lett. **86**, 3200–3203 (2001)
11. Goldenberg, J., Libai, B., Muller, E.: Talk of the network: a complex systems look at the underlying process of word-of-mouth. Mark. Lett. **12**(3), 211–223 (2001)
12. Cohen, R., Erez, K., ben Avraham, D., Havlin, S.: Resilience of the internet to random breakdowns. Phys. Rev. Lett. **85**, 4626–4628 (2000)
13. Li, D., Fu, B., Wang, Y., Lu, G., Berezin, Y., Stanley, H.E., Havlin, S.: Percolation transition in dynamical traffic network with evolving critical bottlenecks. Proc. Natl. Acad. Sci. **1123**(3), 669–672 (2015)
14. Yamasaki, K., Gozolchiani, A., Havlin, S.: Climate networks around the globe are significantly affected by El Niño. Phys. Rev. Lett. **100**, 228501 (2008)
15. Ludescher, J., Gozolchiani, A., Bogachev, M.I., Bunde, A., Havlin, S., Schellnhuber, H.J.: Very early warning of next El Niño. Proc. Natl. Acad. Sci. **111**(6), 2064–2066 (2014)
16. Bunde, A., Havlin, S.: Fractals and Disordered Systems. Springer, New York (1991)
17. Stauffer, D., Aharony, A.: Introduction To Percolation Theory. Taylor & Francis, London/Bristol (1994)
18. Goldenfeld, N.: Lectures on Phase Transitions and the Renormalization Group. Frontiers in Physics. Addison-Wesley, Advanced Book Program, Reading (1992)
19. Albert, R., Jeong, H., Barabási, A.L.: Error and attack tolerance of complex networks. Nature **406**(6794), 378–382 (2000)
20. Barthélemy, M.: Spatial networks. Phys. Rep. **499**(1–3), 1–101 (2011)
21. Bianconi, G.: Statistical mechanics of multiplex networks: entropy and overlap. Phys. Rev. E **87**, 062806 (2013)
22. Nicosia, V., Bianconi, G., Latora, V., Barthelemy, M.: Growing multiplex networks. Phys. Rev. Lett. **111**, 058701 (2013)
23. De Domenico, M., Solé-Ribalta, A., Cozzo, E., Kivelä, M., Moreno, Y., Porter, M.A., Gómez, S., Arenas, A.: Mathematical formulation of multilayer networks. Phys. Rev. X **3**, 041022 (2013)
24. Kivelä, M., Arenas, A., Barthelemy, M., Gleeson, J.P., Moreno, Y., Porter, M.A.: Multilayer networks. J. Complex Netw. **2**(3), 203–271 (2014)
25. Goldenberg, J., Shavitt, Y., Shir, E., Solomon, S.: Distributive immunization of networks against viruses using the 'honey-pot' architecture dimension of spatially embedded networks. Nat. Phys. **1**(3), 184–188 (2005)
26. Rinaldi, S., Peerenboom, J., Kelly, T.: Identifying, understanding, and analyzing critical infrastructure interdependencies. Control Syst. IEEE **21**(6), 11–25 (2001)

27. Hokstad, P., Utne, I., Vatn, J.: Risk and Interdependencies in Critical Infrastructures: A Guideline for Analysis. Springer Series in Reliability Engineering. Springer, London (2012)
28. Buldyrev, S.V., Parshani, R., Paul, G., Stanley, H.E., Havlin, S.: Catastrophic cascade of failures in interdependent networks. Nature **464**(7291), 1025–1028 (2010)
29. Danziger, M.M., Bashan, A., Berezin, Y., Shekhtman, L.M., Havlin, S.: An introduction to interdependent networks. In: Mladenov, V., Ivanov, P. (eds.) Nonlinear Dynamics of Electronic Systems. Volume 438 of Communications in Computer and Information Science, pp. 189–202. Springer, Cham (2014)
30. Parshani, R., Buldyrev, S.V., Havlin, S.: Interdependent networks: reducing the coupling strength leads to a change from a first to second order percolation transition. Phys. Rev. Lett. **105**, 048701 (2010)
31. Gao, J., Buldyrev, S.V., Havlin, S., Stanley, H.E.: Robustness of a network of networks. Phys. Rev. Lett. **107**, 195701 (2011)
32. Kenett, D.Y., Gao, J., Huang, X., Shao, S., Vodenska, I., Buldyrev, S.V., Paul, G., Stanley, H.E., Havlin, S.: Network of interdependent networks: overview of theory and applications. In: D'Agostino, G., Scala, A. (eds.) Networks of Networks: The Last Frontier of Complexity. Understanding Complex Systems, pp. 3–36. Springer, Cham (2014)
33. Foster Jr, J.S., Gjelde, E., Graham, W.R., Hermann, R.J., Kluepfel, H.M., Lawson, R.L., Soper, G.K., Wood, L.L., Woodard, J.B.: Report of the commission to assess the threat to the united states from electromagnetic pulse (emp) attack: critical national infrastructures. Technical report, DTIC Document (2008)
34. Bashan, A., Berezin, Y., Buldyrev, S.V., Havlin, S.: The extreme vulnerability of interdependent spatially embedded networks. Nat. Phys. **9**, 667–672 (2013)
35. D'Agostino, G., Scala, A.: Networks of Networks: The Last Frontier of Complexity. Understanding Complex Systems. Springer, Cham (2014)
36. Boccaletti, S., Bianconi, G., Criado, R., Del Genio, C., Gómez-Gardeñes, J., Romance, M., Sendina-Nadal, I., Wang, Z., Zanin, M.: The structure and dynamics of multilayer networks. Phys. Rep. **544**, 1–122 (2014)
37. Motter, A.E.: Cascade control and defense in complex networks. Phys. Rev. Lett. **93**, 098701 (2004)
38. Dobson, I., Carreras, B.A., Lynch, V.E., Newman, D.E.: Complex systems analysis of series of blackouts: cascading failure, critical points, and self-organization. Chaos: Interdisc. J. Nonlinear Sci. **17**(2), 026103 (2007)
39. Baxter, G.J., Dorogovtsev, S.N., Goltsev, A.V., Mendes, J.F.F.: Avalanche collapse of interdependent networks. Phys. Rev. Lett. **109**, 248701 (2012)
40. Zhou, D., Bashan, A., Cohen, R., Berezin, Y., Shnerb, N., Havlin, S.: Simultaneous first- and second-order percolation transitions in interdependent networks. Phys. Rev. E **90**, 012803 (2014)
41. Newman, M.E.J., Strogatz, S.H., Watts, D.J.: Random graphs with arbitrary degree distributions and their applications. Phys. Rev. E **64**, 026118 (2001)
42. Watanabe, S., Kabashima, Y.: Cavity-based robustness analysis of interdependent networks: influences of intranetwork and internetwork degree-degree correlations. Phys. Rev. E **89**, 012808 (2014)
43. Zhou, D., Gao, J., Stanley, H.E., Havlin, S.: Percolation of partially interdependent scale-free networks. Phys. Rev. E **87**, 052812 (2013)
44. Leicht, E.A., D'Souza, R.M.: Percolation on interacting networks. ArXiv e-prints (2009). http://arxiv.org/abs/0907.0894
45. Hu, Y., Ksherim, B., Cohen, R., Havlin, S.: Percolation in interdependent and interconnected networks: abrupt change from second- to first-order transitions. Phys. Rev. E **84**, 066116 (2011)
46. Parshani, R., Buldyrev, S.V., Havlin, S.: Critical effect of dependency groups on the function of networks. Proc. Natl. Acad. Sci. **108**(3), 1007–1010 (2011)
47. Bashan, A., Parshani, R., Havlin, S.: Percolation in networks composed of connectivity and dependency links. Phys. Rev. E **83**, 051127 (2011)

48. Zhao, J.H., Zhou, H.J., Liu, Y.Y.: Inducing effect on the percolation transition in complex networks. Nat. Commun. **4**, 2412 (2013)
49. Shao, J., Buldyrev, S.V., Havlin, S., Stanley, H.E.: Cascade of failures in coupled network systems with multiple support-dependence relations. Phys. Rev. E **83**, 036116 (2011)
50. Gao, J., Buldyrev, S.V., Stanley, H.E., Havlin, S.: Networks formed from interdependent networks. Nat. Phys. **8**(1), 40–48 (2012)
51. Gao, J., Buldyrev, S.V., Stanley, H.E., Xu, X., Havlin, S.: Percolation of a general network of networks. Phys. Rev. E **88**, 062816 (2013)
52. Gao, J., Buldyrev, S.V., Havlin, S., Stanley, H.E.: Robustness of a network formed by n interdependent networks with a one-to-one correspondence of dependent nodes. Phys. Rev. E **85**, 066134 (2012)
53. Zhou, D., Stanley, H.E., D'Agostino, G., Scala, A.: Assortativity decreases the robustness of interdependent networks. Phys. Rev. E **86**, 066103 (2012)
54. Parshani, R., Rozenblat, C., Ietri, D., Ducruet, C., Havlin, S.: Inter-similarity between coupled networks. EPL (Europhys. Lett.) **92**(6), 68002 (2010)
55. Buldyrev, S.V., Shere, N.W., Cwilich, G.A.: Interdependent networks with identical degrees of mutually dependent nodes. Phys. Rev. E **83**, 016112 (2011)
56. Valdez, L.D., Macri, P.A., Stanley, H.E., Braunstein, L.A.: Triple point in correlated interdependent networks. Phys. Rev. E **88**, 050803 (2013)
57. Lee, K.M., Kim, J.Y., Cho, W.k., Goh, K.I., Kim, I.M.: Correlated multiplexity and connectivity of multiplex random networks. New J. Phys. **14**(3), 033027 (2012)
58. Cellai, D., López, E., Zhou, J., Gleeson, J.P., Bianconi, G.: Percolation in multiplex networks with overlap. Phys. Rev. E **88**, 052811 (2013)
59. Li, M., Liu, R.R., Jia, C.X., Wang, B.H.: Critical effects of overlapping of connectivity and dependence links on percolation of networks. New J. Phys. **15**(9), 093013 (2013)
60. Hu, Y., Zhou, D., Zhang, R., Han, Z., Rozenblat, C., Havlin, S.: Percolation of interdependent networks with intersimilarity. Phys. Rev. E **88**, 052805 (2013)
61. Newman, M.E.J.: Random graphs with clustering. Phys. Rev. Lett. **103**, 058701 (2009)
62. Huang, X., Shao, S., Wang, H., Buldyrev, S.V., Eugene Stanley, H., Havlin, S.: The robustness of interdependent clustered networks. EPL **101**(1), 18002 (2013)
63. Shao, S., Huang, X., Stanley, H.E., Havlin, S.: Robustness of a partially interdependent network formed of clustered networks. Phys. Rev. E **89**, 032812 (2014)
64. Shekhtman, L.M., Berezin, Y., Danziger, M.M., Havlin, S.: Robustness of a network formed of spatially embedded networks. Phys. Rev. E **90**, 012809 (2014)
65. Gao, J., Li, D., Havlin, S.: From a single network to a network of networks. Natl. Sci. Rev. **1**(3), 346–356 (2014)
66. Danziger, M.M., Bashan, A., Berezin, Y., Havlin, S.: Percolation and cascade dynamics of spatial networks with partial dependency. J. Complex Netw. **2**, 460–474 (2014)
67. Rosato, V., Issacharoff, L., Tiriticco, F., Meloni, S., Porcellinis, S.D., Setola, R.: Modelling interdependent infrastructures using interacting dynamical models. Int. J. Crit. Infrastruct. **4**(1/2), 63 (2008)
68. Hines, P., Blumsack, S., Cotilla Sanchez, E., Barrows, C.: The topological and electrical structure of power grids. In: 2010 43rd Hawaii International Conference on System Sciences (HICSS), Koloa, pp. 1–10 (2010)
69. Li, D., Kosmidis, K., Bunde, A., Havlin, S.: Dimension of spatially embedded networks. Nat. Phys. **7**(6), 481–484 (2011)
70. Li, W., Bashan, A., Buldyrev, S.V., Stanley, H.E., Havlin, S.: Cascading failures in interdependent lattice networks: the critical role of the length of dependency links. Phys. Rev. Lett. **108**, 228702 (2012)
71. Berezin, Y., Bashan, A., Danziger, M.M., Li, D., Havlin, S.: Localized attacks on spatially embedded networks with dependencies. Sci. Rep. **5**, 8934 (2015)
72. Danziger, M.M., Bashan, A., Berezin, Y., Havlin, S.: Interdependent spatially embedded networks: dynamics at percolation threshold. In: 2013 International Conference on Signal-Image Technology Internet-Based Systems (SITIS), Kyoto, pp. 619–625 (2013)

73. Nienhuis, B.: Analytical calculation of two leading exponents of the dilute potts model. J. Phys. A: Math. Gen. **15**(1), 199 (1982)
74. Danziger, Michael M., Bashan, Amir, Havlin, Shlomo: Interdependent resistor networks with process-based dependency. New J. Phys. **17**(4), 043046 (2015)
75. Danziger, M.M., Shekhtman, L.M., Berezin, Y., Havlin, S.: Two distinct transitions in spatially embedded multiplex networks. ArXiv e-prints (2015). http://arxiv.org/abs/1505.01688
76. Huang, X., Gao, J., Buldyrev, S.V., Havlin, S., Stanley, H.E.: Robustness of interdependent networks under targeted attack. Phys. Rev. E **83**, 065101 (2011)
77. Dong, G., Gao, J., Du, R., Tian, L., Stanley, H.E., Havlin, S.: Robustness of network of networks under targeted attack. Phys. Rev. E **87**, 052804 (2013)
78. Schneider, C.M., Yazdani, N., Araújo, N.A., Havlin, S., Herrmann, H.J.: Towards designing robust coupled networks. Sci. Rep. **3**, 1969 (2013)
79. Valdez, L.D., Macri, P.A., Braunstein, L.A.: A triple point induced by targeted autonomization on interdependent scale-free networks. J. Phys. A: Math. Theor. **47**(5), 055002 (2014)
80. Holland, P.W., Laskey, K.B., Leinhardt, S.: Stochastic blockmodels: first steps. Soc. Netw. **5**(2), 109–137 (1983)
81. Karrer, B., Newman, M.E.J.: Stochastic blockmodels and community structure in networks. Phys. Rev. E **83**, 016107 (2011)
82. Peixoto, T.P., Bornholdt, S.: Evolution of robust network topologies: emergence of central backbones. Phys. Rev. Lett. **109**, 118703 (2012)
83. Stippinger, M., Kertész, J.: Enhancing resilience of interdependent networks by healing. Phys. A: Stat. Mech. Appl. **416**(0), 481–487 (2014)
84. Agarwal, P.K., Efrat, A., Ganjugunte, S., Hay, D., Sankararaman, S., Zussman, G.: The resilience of WDM networks to probabilistic geographical failures. In: 2011 Proceedings IEEE INFOCOM, Shanghai, pp. 1521–1529 (2011)
85. Bernstein, A., Bienstock, D., Hay, D., Uzunoglu, M., Zussman, G.: Sensitivity analysis of the power grid vulnerability to large-scale cascading failures. SIGMETRICS Perform. Eval. Rev. **40**(3), 33–37 (2012)
86. Shao, S., Huang, X., Stanley, H.E., Havlin, S.: Percolation of localized attack on complex networks. New J. Phys. **17**(2), 023049 (2015)
87. Son, S.W., Bizhani, G., Christensen, C., Grassberger, P., Paczuski, M.: Percolation theory on interdependent networks based on epidemic spreading. EPL (Europhys. Lett.) **97**(1), 16006 (2012)
88. Saumell-Mendiola, A., Serrano, M.Á., Boguñá, M.: Epidemic spreading on interconnected networks. Phys. Rev. E **86**, 026106 (2012)
89. Dickison, M., Havlin, S., Stanley, H.E.: Epidemics on interconnected networks. Phys. Rev. E **85**, 066109 (2012)
90. Wang, H., Li, Q., D'Agostino, G., Havlin, S., Stanley, H.E., Van Mieghem, P.: Effect of the interconnected network structure on the epidemic threshold. Phys. Rev. E **88**, 022801 (2013)
91. Erez, T., Hohnisch, M., Solomon, S.: Statistical economics on multi-variable layered networks. In: Salzano, M., Kirman, A. (eds.) Economics: Complex Windows. New Economic Windows, pp. 201–217. Springer, Milan (2005)
92. Huang, X., Vodenska, I., Havlin, S., Stanley, H.E.: Cascading failures in bi-partite graphs: model for systemic risk propagation. Sci. Rep. **3**, 1219 (2013)
93. Li, W., Kenett, D.Y., Yamasaki, K., Stanley, H.E., Havlin, S.: Ranking the economic importance of countries and industries. ArXiv e-prints (2014). http://arxiv.org/abs/1408.0443
94. Bashan, A., Bartsch, R.P., Kantelhardt, J.W., Havlin, S., Ivanov, P.C.: Network physiology reveals relations between network topology and physiological function. Nat. Commun. **3**, 702 (2012)
95. Pocock, M.J.O., Evans, D.M., Memmott, J.: The robustness and restoration of a network of ecological networks. Science **335**(6071), 973–977 (2012)
96. Donges, J., Schultz, H., Marwan, N., Zou, Y., Kurths, J.: Investigating the topology of interacting networks. Eur. Phys. J. B **84**(4), 635–651 (2011)

97. Reis, S.D.S., Hu, Y., Babino, A., Andrade Jr, J.S., Canals, S., Sigman, M., Makse, H.A.: Avoiding catastrophic failure in correlated networks of networks. Nat. Phys. **10**(10), 762–767 (2014)
98. Morris, R.G., Barthelemy, M.: Transport on coupled spatial networks. Phys. Rev. Lett. **109**, 128703 (2012)
99. Majdandzic, A., Braunstein, L.A., Curme, C., Vodenska, I., Levy-Carciente, S., Stanley, H.E., Havlin, S.: Multiple tipping points and optimal repairing in interacting networks. ArXiv e-prints (2015). http://arxiv.org/abs/1502.00244

Chapter 6
A Unified Approach to Percolation Processes on Multiplex Networks

Gareth J. Baxter, Davide Cellai, Sergey N. Dorogovtsev, Alexander V. Goltsev, and José F.F. Mendes

Abstract Many real complex systems cannot be represented by a single network, but due to multiple sub-systems and types of interactions, must be represented as a multiplex network. This is a set of nodes which exist in several layers, with each layer having its own kind of edges, represented by different colors. An important fundamental structural feature of networks is their resilience to damage, the percolation transition. Generalization of these concepts to multiplex networks requires careful definition of what we mean by connected clusters. We consider two different definitions. One, a rigorous generalization of the single-layer definition leads to a strong non-local rule, and results in a dramatic change in the response of the system to damage. The giant component collapses discontinuously in a hybrid transition characterized by avalanches of diverging mean size. We also consider another definition, which imposes weaker conditions on percolation and allows local calculation, and also leads to different sized giant components depending on whether we consider an activation or pruning process. This 'weak' process exhibits both continuous and discontinuous transitions.

G.J. Baxter (✉) • J.F.F. Mendes
Department of Physics & I3N, University of Aveiro, Aveiro, Portugal
e-mail: gjbaxter@ua.pt; jfmendes@ua.pt

D. Cellai
MACSI, Department of Mathematics and Statistics, University of Limerick, Limerick, Ireland
e-mail: davide.cellai@ul.ie

S.N. Dorogovtsev • A.V. Goltsev
Department of Physics & I3N, University of Aveiro, Aveiro, Portugal

A. F. Ioffe Physico-Technical Institute, 194021 St. Petersburg, Russia
e-mail: sdorogov@ua.pt; goltsev@ua.pt

© Springer International Publishing Switzerland 2016
A. Garas (ed.), *Interconnected Networks*, Understanding Complex Systems,
DOI 10.1007/978-3-319-23947-7_6

6.1   Introduction

Networks are a powerful tool to represent the heterogeneous structure of interactions in the study of complex systems [12]. But in many cases there are multiple kinds of interactions, or multiple interacting sub-systems that cannot be adequately represented by a single network. Examples include financial [8, 15], infrastructure [21], informatic [16] and ecological [19] systems.

There are many representations of multi-layer networks, appropriate in different circumstances, see [6] for a recent review of the topic. We focus on multiplex networks, which are networks with a single set of nodes present in all layers, connected by a different type of edge (which may be represented by different colors) in each layer. Some interdependent networks, in which different layers have different sets of nodes, but the nodes are connected between layers by interdependency links [7, 14], can be captured by this construction [22].

One of the fundamental structural properties of a network is its response to damage, that is, the percolation transition, where the giant connected component collapses. In multi-layer networks, interdependencies between layers can make a system more fragile. Damage to one element can trigger avalanches of failures that spread through the whole system [13, 20]. Typically a discontinuous hybrid phase transition is observed [3], similar to those observed in the network k-core or in bootstrap percolation [2] in contrast to the continuous transition seen in classical percolation on a simplex network.

Under a weaker definition of percolation, a more complex phase diagram emerges, with the possibility for both continuous and discontinuous transitions. When invulnerable or seed nodes are introduced, we can define activation and pruning processes, which have different phase diagrams. The results presented in this Chapter are based on those obtained in [3] and [4].

In a single-layer network (simplex), two nodes are connected if there is at least one path between them along the edges of the network. A group of connected nodes forms a cluster. The giant connected component (GCC) is a cluster which contains a finite fraction of the nodes in the network. The existence of such a giant component is synonymous with percolation. We can study its appearance by applying random damage to the network. A fraction $1 - p$ of nodes are removed, independently at random, and we check whether the remaining network contains a giant connected component. Typically the GCC appears linearly with a continuous second-order transition, although when the degree distribution is very broad (as in scale-free networks) the nature and location of the transition may be dramatically altered [10].

For multiplex networks, we must generalize these definitions of clusters and percolation. Consider a multiplex network, with nodes $i = 1, 2, \ldots, N$ connected by m colors of edges labeled $s = a, b, \ldots, m$. Two nodes i and j are m-connected if for each of the m types of edges, there is a path from i to j following edges only of that type. Let us suppose that the connections are essential to the function of

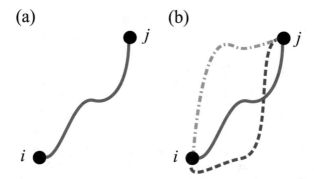

Fig. 6.1 (**a**) In an ordinary network, two vertices i and j belong to the same cluster if there is a path connecting them. (**b**) In a multiplex network, vertices i and j belong to the same viable cluster if there is a path connecting them for every kind of edge, following only edges of that kind. In the example shown, there are $m = 3$ kinds of edges. Vertices i and j are said to be 3-connected (After [5])

each site, so that a node is only viable if it maintains connections of every type to other viable vertices. A viable cluster is then a cluster of m-connected nodes. This definition is described in Fig. 6.1.

In a large system, we wish to find when there is a giant cluster of viable nodes. From this definition of viable clusters, it follows that any giant viable cluster is a subgraph of the giant connected component of each of the m layers formed by considering only a single color of edge in the multiplex network. The absence of a giant connected component in any one of the layers means the absence of the giant viable cluster. Note that when $m = 1$, the viable clusters are identical to clusters of connected vertices in ordinary networks with a single type of edges. As we will see, the rigorous requirements for viability in multiplex networks have a profound effect on the percolation of the network, revealing a discontinuous hybrid transition in the collapse of the giant viable component.

The giant viable cluster is related to the so-called giant mutually connected component in interdependent networks [7, 14]. By definition, a node belongs to the giant mutually connected component if at least one of its neighbors within its own network and its interdependent neighbors in the other network (if it exists) belong to this component. In this way, the giant viable cluster corresponds to the giant mutually connected component in the case of full interdependency, i.e. when each node in one network has a single interdependent replica node in the other net.

If we relax the criterion that a cluster must be connected by all layers, instead requiring only connection via paths of any color of mixture of colors, we immediately return to ordinary percolation, equivalent to projecting all the layers of the multiplex onto a single layer, that is, ignoring all the colors. If, rather, we were to consider clusters of nodes in which each pair is connected by at least one single colored path, the resulting giant connected component would be the union of the connected components of the individual layers.

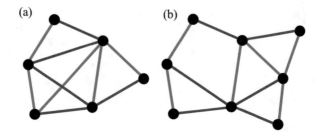

Fig. 6.2 Examples of small connected clusters in the strong and weak definitions of connectedness in a two-layer multiplex network. (**a**) In the strong definition of a cluster, every node in a viable cluster can reach every other by every kind of edge. (**b**) In the weak definition, every node has connections of both colors, but there is not necessarily a path of every color between every pair of nodes

Instead, we may consider a more interesting definition, which is still weaker than the viable clusters defined above. To differentiate it from the definition above, we will call this *weak* percolation. We continue with the requirement that each node only functions if it is connected to other functioning nodes by every color of edge. However, it does not need to be connected to every node in the cluster by every kind of edge. Weak percolation can be defined in the following way: a node i is active if, for each of the m colors, it is connected to at least one active neighbor by an edge of that color. Weak percolating clusters are then simply connected clusters of active nodes. Examples of the connected clusters resulting from the two different rules can be compared in Fig. 6.2.

We can consider an activation process, in which a small number of nodes are initially activated, and activation may spread to neighboring nodes. This can represent, for example, social mobilization or the repair of infrastructure after a disaster [13]. This generalizes activation processes such as bootstrap percolation [1] to multiplex networks. Comparing with the counterpart pruning process, we find that the two processes do not result in the same giant active component [4]. A similar problem was considered in [17].

In the following Section, we analyze the strong definition of percolation on multiplex networks, identifying the nature of the percolation transition and the associated avalanches of damage. In Sect. 6.3, we analyze the weak definition of percolation and explore the activation and pruning processes, showing that they also exhibit hybrid transitions, and outlining the complex phase diagrams that appear.

6.2 Multiplex Percolation

The viable clusters in a multiplex network can be identified by an iterative pruning process, testing the connectivity in every layer, and removing nodes that fail. Such removals may affect the connectivity of the remaining nodes, so we must repeat the

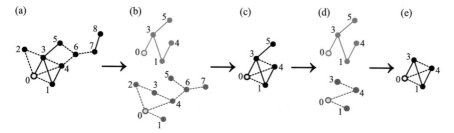

Fig. 6.3 An example demonstrating the algorithm for identifying a viable cluster in a small network with two kinds of edges. (**a**) In the original network, in step (i) we select vertex 0 as the test vertex. (**b**) In step (ii) we identify the clusters of vertices connected to 0 by each kind of edge. (**c**) Step (iii): the intersection of these two clusters becomes the new candidate set for the viable cluster to which 0 belongs. (**d**) We repeat steps (ii) using only vertices from the candidate set shown in (**c**). Repeating step (iii), we find the overlap between the two clusters from (**d**), shown in (**e**). Further repetition of steps (ii) and (iii) does not change this cluster, meaning that the cluster consisting of vertices 0, 1, 3 and 4 is a viable cluster (After [5])

process until an equilibrium is reached. An algorithm for identifying viable clusters is the following:

(i) Choose a test vertex i at random from the network.
(ii) For each kind of edge s, compile a list of vertices that can be reached from i by following only edges of type s.
(iii) The intersection of these m lists forms a new candidate set for the viable cluster containing i.
(iv) Repeat steps (ii) and (iii) but traversing only the current candidate set. When the candidate set no longer changes, it is either a viable cluster, or contains only vertex i.
(v) To find further viable clusters, remove the viable cluster of i from the network (cutting any edges) and repeat steps (i)–(iv) on the remaining network beginning from a new test vertex.

Note that this process is non-local: it is not possible to identify whether a node is a member of a viable cluster simply by examining its immediate neighbors. An example of the use of this algorithm in a small network is illustrated in Fig. 6.3.

We now study in more detail the collapse of the giant viable cluster under damage by random removal of nodes. We use the fraction p of undamaged nodes as a control variable. In uncorrelated random networks the giant viable cluster collapses at a critical undamaged fraction p_c in a discontinuous hybrid transition, similar to that seen in the k-core or bootstrap percolation [1, 11].

Let us consider sparse uncorrelated networks, which are locally tree-like in the infinite size limit $N \to \infty$. We take advantage of this locally tree-like property to define recursive equations which allow us to find the giant viable cluster. We define X_s, with the index $s \in \{a, b, \ldots\}$, to be the probability that, on following an arbitrarily chosen edge of type s, we encounter the root of an infinite sub-tree formed solely from type s edges, whose vertices are also each connected to at least

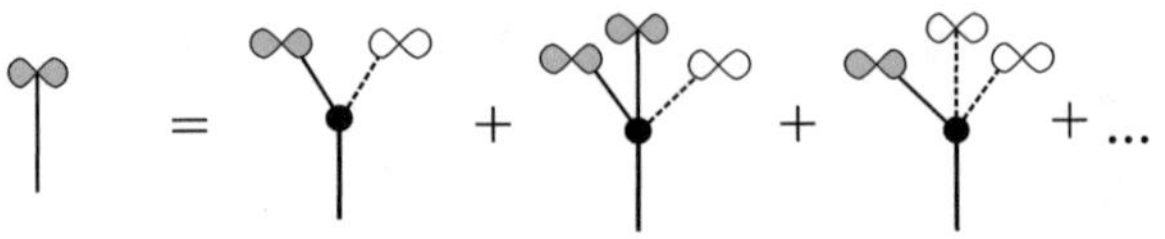

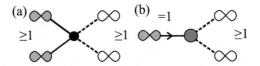

Fig. 6.4 Diagrammatic representation of Eq. (6.1) in a system of two interdependent networks a and b. The probability X_a, represented by a *shaded infinity symbol* can be written recursively as a sum of second-neighbor probabilities. *Open infinity symbols* represent the equivalent probability X_b for network b, which obeys a similar recursive equation. The *filled circle* represents the probability p that the vertex remains in the network (After [3])

Fig. 6.5 Viable and critical viable vertices for two interdependent networks. (**a**) A vertex is in the giant viable cluster if it has connections of both kinds to giant viable subtrees, represented by *infinity symbols*, which occur with probabilities X_a (*shaded*) or X_b (*open*) – see text. (**b**) A critical viable vertex of type a has exactly 1 connection to a giant sub-tree of type a (After [3])

one infinite subtree of every other type. We call this a type s infinite subtree. The vector $\{X_a, X_b, \ldots\}$ plays the role of the order parameter. In a two-layer network, for example, the probability X_a can be written as the sum of second-level probabilities in terms of X_a and X_b, as illustrated in Fig. 6.4. In general, writing this graphical representation in equation form, using the joint degree distribution $P(q_a, q_b, \ldots)$, we arrive at the self consistency equations (for more details, see [3])

$$X_s = p \sum_{q_a, q_b, \ldots} \frac{q_s}{\langle q_s \rangle} P(q_a, q_b, \ldots) \left[1 - (1 - X_s)^{q_s - 1}\right] \prod_{l \neq s} \left[1 - (1 - X_l)^{q_l}\right]$$

$$\equiv \Psi_s(X_a, X_b, \ldots), \tag{6.1}$$

where p is the probability that the vertex was not initially damaged. The term $(q_s/\langle q_s \rangle)P(q_a, q_b, \ldots)$ gives the probability that on following an arbitrary edge of type s, we find a vertex with degrees $q_a, q_b, \ldots$, while $[1 - (1 - X_a)^{q_a}]$ is the probability that this vertex has at least one edge of type $a \neq s$ leading to the root of an infinite sub-tree of type a edges. This becomes $[1 - (1 - X_s)^{q_s - 1}]$ when $a = s$.

A vertex is then in the giant viable cluster if it has at least one edge of every type s leading to an infinite type s sub-tree (probability X_s), as shown in Fig. 6.5a.

$$S = p \sum_{q_a, q_b, \ldots} P(q_a, q_b, \ldots) \prod_{s = a, b, \ldots} \left[1 - (1 - X_s)^{q_s}\right], \tag{6.2}$$

which is equal to the relative size of the giant viable cluster of the damaged network.

A hybrid transition appears at the point where $\Psi_s(X_a, X_b, \ldots)$ first meets X_s at a non-zero value, for all s. This occurs when

$$\det[\mathbf{J} - \mathbf{I}] = 0, \tag{6.3}$$

where $\mathbf{I}$ is the unit matrix and $\mathbf{J}$ is the Jacobian matrix $J_{ab} = \partial\Psi_b/\partial X_a$. The critical point p_c can then be found by simultaneously solving Eqs. (6.1) and (6.3). To find the scaling near the critical point, we expand Eq. (6.1) about the critical value $X_s^{(c)}$. We find that

$$X_s - X_s^{(c)} \propto (p - p_c)^{1/2}. \tag{6.4}$$

This square-root scaling is the typical behavior of the order parameter near a hybrid transition. It results from avalanches of spreading damage which diverge in size near the transition. The scaling of the size of the giant viable cluster, S, immediately follows

$$S - S_c \propto (p - p_c)^{1/2}. \tag{6.5}$$

Avalanches

We now examine the avalanches of damage which occur in the system, in order to understand the nature of the transition more completely. We focus on the case of two types of edges. Consider a viable node that has exactly one edge of type a leading to a type a infinite subtree, and at least one edge of type b leading to a type b infinite subtree. We call this a critical node of type a. It is illustrated in Fig. 6.5b. It is a critical vertex because it will be removed from the viable cluster if it loses its single link to a type a infinite subtree. The removal of any node from the giant viable cluster, and the edges to which it is connected, therefore also requires the removal of any critical vertices which depend on the removed edges. Removed critical nodes may have edges leading to further critical nodes. This is the way that damage propagates in the system. The removal of a single node can result in an avalanche of removals of critical vertices from the giant viable cluster.

To represent this process visually, we draw a diagram of viable nodes and the edges between them. We mark the special critical edges, that critical viable nodes depend on, with an arrow leading to the critical node. An avalanche can only transmit in the direction of the arrows. For example, in Fig. 6.6, removal of the vertex labeled 1 removes the essential edge of the critical vertex 2 which thus becomes non-viable. Removal of vertex 2 causes the removal of further critical vertices 3 and 4, and the removal of 4 then requires the removal of 5. Thus critical vertices form critical clusters. Graphically, upon removal of a vertex, we remove all vertices found by following the arrowed edges, which constitutes an avalanche. Note that

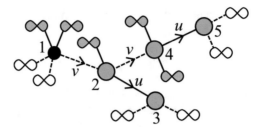

Fig. 6.6 A critical cluster. Removal of any of the shown viable vertices will result in the removal of all downstream critical viable vertices. Vertices 2–5 are critical vertices. Removal of the vertex labeled 1 will result in all of the shown vertices being removed (becoming non-viable). Removal of vertex 2 results in the removal of vertices 3, 4, and 5 as well, while removal of vertex 4 results only in vertex 5 also being removed. As before, infinity symbols represent connections to infinite viable subtrees. Other connections to non-viable vertices or finite viable clusters are not shown (After [3])

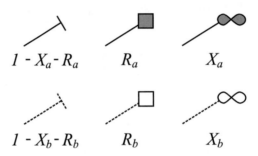

Fig. 6.7 Symbols used in the diagrams to represent key probabilities. *Solid lines* represent edges of type *a*, *dashed lines* represent edges of type *b* (After [5])

an avalanche is a branching process. Removing a vertex may lead to avalanches along several edges emanating from the vertex (for example, in Fig. 6.6, removing vertex 2 leads to avalanches along two edges). As we approach the critical point from above, the avalanches increase in size. The mean size of avalanches triggered by a randomly removed vertex finally diverges in size at the critical point, which is the cause of the discontinuity in the size of the giant viable cluster, which collapses to zero. These avalanches are thus an inherent part of a hybrid transition. To show this, we use a generating function approach [18] to calculate the sizes and structure of avalanches.

There are three possibilities when following an arbitrarily chosen edge of a given type: (i) with probability X_s we encounter a type s infinite subtree (ii) with probability R_s we encounter a vertex which has a connection to an infinite subtree of the opposite type, but none of the same type. Such a vertex is part of the giant viable cluster if the parent vertex was; or (iii) with probability $1 - X_s - R_s$, we encounter a vertex which has no connections to infinite subtrees of either kind. These probabilities are represented graphically in Fig. 6.7. We will use these symbols in subsequent diagrams.

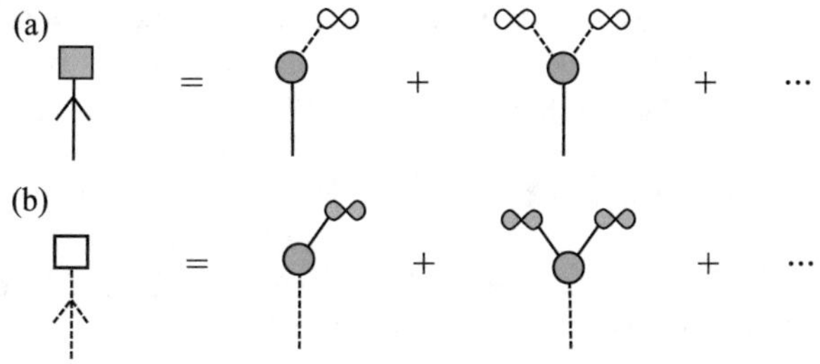

Fig. 6.8 (a) The probability R_a can be defined in terms of the second-level connections of the vertex found upon following an edge of type a. Note that possible connections to 'dead ends' – vertices not in the viable cluster (probability $1 - X_a - R_a$ or $1 - X_b - R_b$) are not shown. (b) The equivalent graphical equation for the probability R_b (After [5])

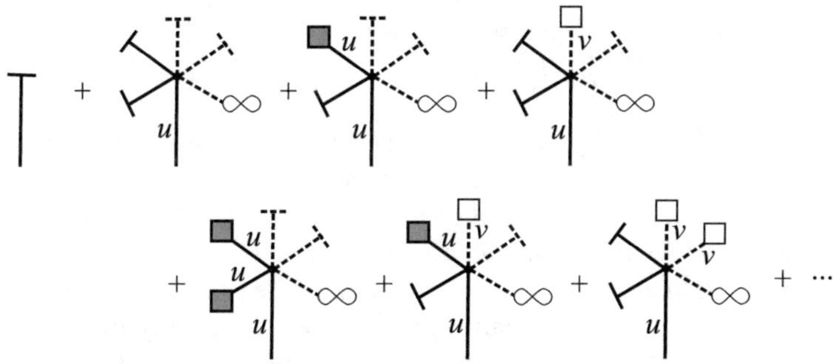

Fig. 6.9 Representation of the generating function $H_a(x, y)$ (right-hand side of Eq. 6.7) for the size of a critical cluster encountered upon following an edge of type a (After [5])

The probability R_a obeys

$$R_a = \sum_{q_a} \sum_{q_b} \frac{q_a}{\langle q_a \rangle} P(q_a, q_b)(1 - X_a)^{q_a-1} \left[1 - (1 - X_b)^{q_b} \right] \tag{6.6}$$

and similarly for R_b. This equation is represented graphically in Fig. 6.8.

The generating function for the size of the avalanche triggered by removing an arbitrary type a edge which does not lead to an infinite type a subtree can be found by considering the terms represented in Fig. 6.9. The first term represents the probability that, upon following an edge of type a (solid lines) we reach a node with no connection to a type b subtree (and hence is not viable), that is, a critical cluster of size 0. The second term represents the probability to encounter a critical cluster of size 1. The node encountered has a connection to the type b infinite subtree (infinity symbol), but no further connections to viable nodes. Subsequent terms represent recursive probabilities that the vertex encountered has 1 (third and fourth terms), 2 (fifth, sixth, seventh terms) or more connections to further potential critical clusters. The variables u (for type a edges) and v (type b) are assigned to each such edges.

Considering these terms, we can write the generating function for the number of critical vertices encountered upon following an arbitrary edge of type a (that is, the size of the resulting avalanche if this edge is removed) as

$$H_a(u, v) = 1 - X_a - R_a + uF_a[H_a(u, v), H_b(u, v)] \tag{6.7}$$

and similarly for $H_b(u, v)$, the corresponding generating function for the size of the avalanche caused by removing a type b edge is

$$H_b(u, v) = 1 - X_b - R_b + vF_b[H_a(u, v), H_b(u, v)], \tag{6.8}$$

where the functions $F_a(x, y)$ and $F_b(x, y)$ are defined as:

$$F_a(x, y) \equiv \sum_{q_a} \sum_{q_b} \frac{q_a}{\langle q_a \rangle} P(q_a, q_b) x^{q_a-1} \sum_{r=1}^{q_b} \binom{q_b}{r} X_b^r y^{q_b-r} \tag{6.9}$$

and similarly for $F_b(x, y)$, by exchanging all subscripts a and b. While the function $F_a(x, y)$ does not necessarily represent a physical quantity or probability, we can see that it incorporates the probability of encountering a vertex with at least one child edge of type b leading to a giant viable subtree (probability X_b) upon following an edge of type a. All other outgoing edges then contribute a factor x (for type a edges) or y (type b). Here u and v are auxiliary variables. Following through a critical cluster, a factor u appears for each arrowed edge of type a, and v for each arrowed edge of type b. For example, the critical cluster illustrated in Fig. 6.6 contributes a factor $u^2 v^2$.

The mean number of critical vertices reached upon following an edge of type a, i.e. the mean size of the resulting avalanche if this edge is removed, is given by $\partial_u H_a(1, 1) + \partial_v H_a(1, 1)$, where ∂_u signifies the partial derivative with respect to u. Unbounded avalanches emerge at the point where $\partial_u H_a(1, 1)$ [or $\partial_v H_b(1, 1)$] diverges. Taking derivatives of Eq. (6.7),

$$\partial_u H_a(u, v) = F_a[H_a, H_b] + u\{\partial_u H_a \partial_x F_a[H_a, H_b] + \partial_u H_b \partial_y F_a[H_a, H_b]\} \tag{6.10}$$

$$\partial_v H_a(u, v) = u\{\partial_v H_a \partial_x F_a[H_a, H_b] + \partial_v H_b \partial_y F_a[H_a, H_b]\} \tag{6.11}$$

with similar equations for $\partial_u H_b(u, v)$ and $\partial_v H_b(u, v)$. Some rearranging gives

$$\partial_u H_a(1, 1) = \frac{R_a + \partial_u H_b(1, 1)\partial_y F_a(1 - X_a, 1 - X_b)}{1 - \partial_x F_a(1 - X_a, 1 - X_b)} \tag{6.12}$$

and

$$\partial_v H_a(1, 1) = \frac{\partial_u H_a(1, 1)\partial_x F_b(1 - X_a, 1 - X_b)}{1 - \partial_y F_b(1 - X_a, 1 - X_b)} \tag{6.13}$$

where we have used that $H_a(1,1) = 1 - X_a$ and $F_a(1 - X_a, 1 - X_b) = R_a$. From Eqs. (6.1) and (6.9),

$$\partial_x F_a(1 - X_a, 1 - X_b) = \frac{\partial}{\partial X_a} \Psi_a(X_a, X_b) \tag{6.14}$$

$$\partial_y F_b(1 - X_a, 1 - X_b) = \frac{\langle q_a \rangle}{\langle q_b \rangle} \frac{\partial}{\partial X_a} \Psi_b(X_a, X_b), \tag{6.15}$$

and similarly for $\partial_x F_b$ and $\partial_y F_b$, which when substituted into (6.12) and (6.13) give

$$\partial_u H_a(1,1) = \frac{R_a[1 - \frac{\partial}{\partial X_b} \Psi_b(X_a, X_b)]}{\det[\mathbf{J} - \mathbf{I}]}. \tag{6.16}$$

We see that the denominator exactly matches the left-hand side of Eq. (6.3), meaning that the mean size of avalanches triggered by random removal of vertices diverges exactly at the point of the hybrid transition.

6.3 Weak Multiplex Percolation

Now we consider, for comparison, the weaker definition of percolation on multiplex networks. In this case we also find a discontinuous hybrid transition, but a continuous second order transition may also occur.

In ordinary percolation, and the strong multiplex percolation considered above, activation and deactivation yield the same giant cluster. In weak percolation, however, activation of the network yields a very different phase diagram than a pruning process. We define an activation process, which we call Weak Bootstrap Percolation (WBP) and a deactivation/pruning process, Weak Pruning Percolation (WPP). We also introduce invulnerable vertices, which are always active. These are necessary to seed the activation process, and we include them in the pruning process, for symmetry.

Weak Pruning Percolation (WPP)

Let us begin with Weak Pruning Percolation. A fraction f of the nodes are randomly assigned as invulnerable, the rest being vulnerable. In the WPP process, the network is then damaged, with a fraction p of all nodes remaining undamaged. Once again, p acts as a control parameter. Each of the remaining vulnerable nodes is pruned if it fails to have at least one connection in each layer to a surviving node (vulnerable or invulnerable). The removal of some nodes may affect the neighborhoods of other surviving nodes, so the pruning process must be repeated until no more nodes can be removed. Invulnerable nodes cannot be pruned.

Let Z_s be the probability that, upon following an edge of type s, we encounter the root of a sub-tree (whether finite or infinite) formed solely from type s edges, whose vertices are also each connected to at least one such subtree of every other type. We define X_s as the probability that such a subtree is infinite. Precisely, X_s is the probability that each member the subtree encountered, as well as meeting the criteria for Z_s, also has at least one edge leading to an infinite subtree of any type (probability X_a etc.).

In a multiplex with m types of edges and a degree distribution $P(q_a, q_b, \ldots)$, the equation for Z_s is (see [4] for more details):

$$Z_s = pf + p(1-f) \sum_{q_a, q_b, \ldots} \frac{q_s P(q_a, q_b, \ldots)}{\langle q_s \rangle} \prod_{n \neq s} [1 - (1 - Z_n)^{q_n}] \equiv \Phi_s(Z_a, Z_b, \ldots).$$

(6.17)

The first term (pf) accounts for the probability that the encountered node is an undamaged invulnerable node, which is always active, and so its state doesn't depend on any of its neighbors. The second term (proportional to $p(1-f)$) calculates the recursive probability for vulnerable undamaged nodes.

The equation for X_s is

$$X_s = pf \sum_{q_a, q_b, \ldots} \frac{q_s P(q_a, q_b, \ldots)}{\langle q_s \rangle} \left[1 - (1 - X_s)^{q_s - 1} \prod_{n \neq s} (1 - X_n)^{q_n} \right]$$

$$+ p(1-f) \sum_{q_a, q_b, \ldots} \frac{q_s P(q_a, q_b, \ldots)}{\langle q_s \rangle} \left\{ \prod_{n \neq s} [1 - (1 - Z_n)^{q_n}] - (1 - X_s)^{q_s - 1} \right.$$

$$\left. \times \prod_{n \neq s} [(1 - X_n)^{q_n} - (1 - Z_n)^{q_n}] \right\}$$

$$\equiv \Psi_s(X_a, X_b, \ldots, Z_a, Z_b, \ldots)$$.   (6.18)

The first sum on the right hand side calculates the probability that we encounter an undamaged invulnerable node, which has at least one child edge leading to an infinite subtree of any type. The second sum calculates the same probability but in the case when the encountered node is not invulnerable. This term is written as a difference between the probability of having at least one edge leading to finite or infinite subtrees of each type and another term which removes the possibility that all of the subtrees are finite. This last product must be multiplied by $(1 - X_s)^{q_s - 1}$ to exclude the possibility of reaching an infinite subtree by a type s edge.

Finally, given equations for Z_s and X_s, we can use them to find S, i.e. the probability that a randomly chosen node is in the giant percolating cluster defined in this model. This is the strength of the giant percolating cluster. It is given by the following formula:

$$S = pf \sum_{q_a, q_b, \dots} P(q_a, q_b, \dots) \left[1 - \prod_s (1 - X_s)^{q_s} \right]$$

$$+ p(1-f) \sum_{q_a, q_b, \dots} P(q_a, q_b, \dots) \prod_s [1 - (1 - Z_s)^{q_s}] - \prod_s [(1 - X_s)^{q_s} - (1 - Z_s)^{q_s}].$$

(6.19)

This equation is constructed in a similar way to that for X_s.

A continuous transition appears at the point where a non-zero solution to $X_s = \Psi_s$ first appears. A hybrid transition appears at the point where Ψ_s is first tangent to X_s at a non-zero value, for all s. Because a jump in X_s is always accompanied by a jump in Z_s, it is more simple to look for the point where Φ_s is tangent to Z_s. This occurs when

$$\det[\mathbf{J} - \mathbf{I}] = 0, \tag{6.20}$$

where $\mathbf{I}$ is the unit matrix and $\mathbf{J}$ is the Jacobian matrix $J_{ab} = \partial \Phi_b / \partial X_a$. Together these criteria allow us to map the phase diagram of the process with respect to the two parameters f and p.

In the case of only two layers, the phase diagram is characterized by a line of continuous phase transitions. An example is shown in Fig. 6.10, for the case where each of the two layers is an Erdős-Rényi network, with identical mean degree μ. In the limit $f = 0$, the probability of a node being in the giant WPP component is given by the product of the classical percolation probability in each layer. In the Erdős-Rényi example shown in the figure, this means the percolation point is at $\nu \equiv p\mu = 1$. In the limit $f = 1$, all nodes are invulnerable, and the situation

Fig. 6.10 Phase diagram of the WPP model for two uncorrelated Erdős-Rényi networks with identical mean degree μ (After [4])

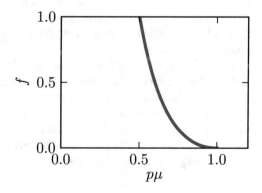

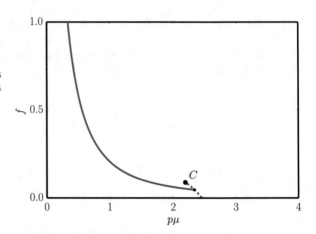

Fig. 6.11 Phase diagram of the WPP model for three uncorrelated Erdős-Rényi networks with identical mean degree μ. The *solid line* gives the location of the continuous transition, the *dashed line* gives the location of the discontinuous transition. The point C is the critical point (After [4])

corresponds to classical percolation with the multiplex is treated as a single network. There is no hybrid transition in the two layer case.

In the case of three layers, now a hybrid transition also appears. The line of discontinuous transitions can be calculated by solving Eqs. (6.17) and (6.20) together. An example phase diagram is given in Fig. 6.11. We see that the both continuous and discontinuous transitions are present, with the giant component appearing discontinuously for small f, and having two transitions for slightly larger f: a continuous appearance followed by a discontinuous hybrid transition.

Weak Bootstrap Percolation (WBP)

Now we consider an activation process called Weak Bootstrap Percolation, which extends the concept of bootstrap percolation [9] to multiplex networks. As for the pruning model, a fraction f of nodes are invulnerable, and are active from the start. Again, a random damage is applied to the network, with the undamaged fraction p acting as the control parameter. Now, however, the vulnerable nodes begin in an inactive state. A node becomes active if it has at least one active neighbor in each of the m layers of the multiplex. The activation of nodes may in turn provide the required active neighbors to more nodes, so the process is repeated until no more nodes can become active.

At the end of the activation process, the active clusters are in general not the same as those that would be found through the pruning process. This is because in WPP nodes are considered active until pruned. This means that, for example, a pair of nodes connected by an edge of one type, provide the required support of that type for one another, even if neither has another edge of that type. In WBP, on the other hand, such an isolated dimer can never become activated (Fig. 6.12). The same holds for many larger configurations as well.

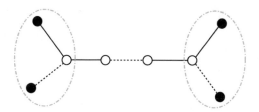

Fig. 6.12 Example of clusters in a multiplex with two types of edges. *Black nodes* are invulnerable/seed vertices, white nodes are vulnerable vertices. In WPP, all the nodes are unprunable (remain active), because each *white node* is connected to another node by each edge type. In WBP, only the nodes inside the *green dot-dashed lines* become active, while the remaining two nodes have only one active neighbor, by one edge type only, so they cannot become active (After [4])

Let Z_s be the probability that, upon following an edge of type s, we encounter the root of a sub-tree (whether finite or infinite) formed solely from type s edges, whose vertices are also each connected to at least one such subtree of every type. This obeys the equation

$$
Z_s = pf + p(1-f) \sum_{q_a, q_b, \dots} \frac{q_s P(q_a, q_b, \dots)}{\langle q_s \rangle} \left[1 - (1 - Z_s)^{q_s - 1} \right] \prod_{n \neq s} [1 - (1 - Z_n)^{q_n}]
$$

$$
\equiv \Phi_s(Z_a, Z_b, \dots).
\tag{6.21}
$$

This differs from the equivalent equation for WPP, Eq. (6.17), because now each node must have connections of every type, not just of the types different from s.

Similarly, we define X_s as the probability that such a subtree is infinite. Precisely, X_s is the probability that each member the subtree encountered, as well as meeting the criteria for Z_s, also has at least one edge leading to an infinite subtree of any type.

An argument similar to the one for Eq. (6.18) leads us to the equation:

$$
X_s = pf \sum_{q_a, q_b, \dots} \frac{q_s P(q_a, q_b, \dots)}{\langle q_s \rangle} \left[1 - (1 - X_s)^{q_s - 1} \prod_{n \neq s} (1 - X_n)^{q_n} \right]
$$

$$
+ p(1-f) \sum_{q_a, q_b, \dots} \frac{q_s P(q_a, q_b, \dots)}{\langle q_s \rangle} \Big\{ [1 - (1 - Z_s)^{q_s - 1}] \prod_{n \neq s} [1 - (1 - Z_n)^{q_n}]
$$

$$
- [(1 - X_s)^{q_s - 1} - (1 - Z_s)^{q_s - 1}] \prod_{n \neq s} [(1 - X_n)^{q_n} - (1 - Z_n)^{q_n}] \Big\}.
\tag{6.22}
$$

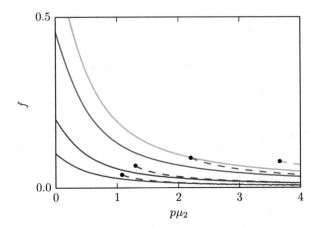

Fig. 6.13 Phase diagram of the WBP model for two uncorrelated Erdős-Rényi networks with mean degree μ_1 and μ_2. *Horizontal axis* is $v_2 = p\mu_2$. Each *solid curve* shows the location of the continuous transition for a particular value of v_1, from *top* to *bottom* $v_1 = \{1.5, 2.193, 5, 10\}$. *Dashed curves* show the corresponding location of the discontinuous transition (which is always above the continuous transition), with *circles marking* the critical end point (After [4])

While Z_n and X_n are different from their WPP counterparts, the equation for S is the same as Eq. (6.19). In the case of WBP, a hybrid transition appears already in a two layer multiplex. A typical phase diagram is plotted in Fig. 6.13, for the case of two Erdős-Rényi layers with different mean degrees. Now we see that the giant component always first appears continuously, with a second, discontinuous hybrid transition occurring afterwards, for small f. The line of discontinuous transitions is obtained using the conditions

$$\begin{cases} \Phi_{f,v_1,v_2}(z) = 1 \\ \Phi'_{f,v_1,v_2}(z) = 0 \end{cases} \tag{6.23}$$

The line ends at the critical point defined by these two conditions in combination with a third condition

$$\Phi''_{f,v_1,v_2}(z) = 0. \tag{6.24}$$

Avalanches

To understand the discontinuous transitions which we observe in the two weak percolation models, we again analyze avalanches, which propagate through clusters of critical vertices. Diverging avalanche sizes lead to the discontinuous transitions. As before, in the pruning process, WPP, a critical vertex is a vertex that just meets

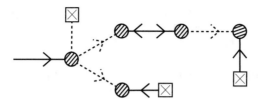

Fig. 6.14 A representation of a cluster of critical vertices in WPP. *Hatching* indicates that vertices are members of the WPP percolating cluster. Because critical vertices are in the percolating cluster for WPP, a critical vertex may be linked to the percolating cluster via another critical vertex. That is, external edges of type Z_s are not necessarily required. Furthermore, this means that critical dependencies can be bi-directional: it is possible for avalanches to propagate in either direction along such edges. Note that outgoing critical edges must be of the opposite type to the incoming one. The *boxes containing crosses* represent the probability $Z_n - R_n$ (After [4])

the criteria for inclusion in the percolating cluster (in the case of WPP). However, in the activation process, WBP, the avalanches which diverge in mean size at the discontinuous transition are of activations of nodes, not of pruning, so critical nodes are those that just fail to meet the criteria for activation.

In the case of WPP, a critical node of type s has exactly one connection to an infinite subtree of type s, and at least one of all the other types. A vertex may be critical with respect to more than one type, if it simultaneously has exactly one connection to infinite subtrees of different types. Such a vertex is related to avalanches because it has one (or possibly more) edge(s) which, if lost, will cause the vertex to be pruned from the cluster. If, in turn, other outgoing edges of this vertex are critical edges for other critical vertices, these vertices will also be removed. Chains of such connections therefore delineate the paths of avalanches of spreading damage. An example is shown in Fig. 6.14. Damage to the node at one end of an edge is transmitted along arrowed edges.

There are three possibilities when following an arbitrarily chosen edge of a given type: (i) with probability X_s we encounter a type s infinite subtree (ii) with probability R_s we encounter a vertex which has a connection to an infinite subtree of the opposite type, but none of the same type. Such a vertex is part of the giant viable cluster if the parent vertex was; or (iii) with probability $1 - X_s - R_s$, we encounter a vertex which has no connections to infinite subtrees of either kind. These probabilities are represented graphically in Fig. 6.7. We will use these symbols in subsequent diagrams.

To examine these avalanches, we define the probability R_s, to be the probability that, on following an edge of type s, we encounter a vulnerable vertex (probability $1 - f$), which has not been removed due to random damage (probability p) and has at least one child edge of each type $n \neq s$ leading to a subtree defined by the probability Z_n, and zero of type s. That is

$$R_s = p(1-f) \sum_{q_a,q_b,\dots} \frac{q_s P(q_a, q_b, \dots)}{\langle q_s \rangle} (1 - Z_s)^{q_s-1} \prod_{n \neq s} [1 - (1 - Z_n)^{q_n}]. \qquad (6.25)$$

We can then define a generating function for the size of the critical subtree encountered upon following an edge of type s (and hence resulting pruning avalanche should the parent vertex of that edge is removed) in a recursive way by

$$H_s(\mathbf{u}) = Z_s - R_s + u_s F_s[H_1(\mathbf{u}), H_2(\mathbf{u}), \ldots, H_m(\mathbf{u})]. \tag{6.26}$$

Where the functions $F_s(\mathbf{x})$ are defined to be

$$F_s(\mathbf{x}) = p(1-f) \sum_{q_a,q_b,\ldots} \frac{q_s P(q_a, q_b, \ldots)}{\langle q_s \rangle} (1-Z_s)^{q_m-1} \prod_{n \neq s} \sum_{l=1}^{q_s} \binom{q_s}{l} (1-Z_n)^{q_n-l} x_n^l.$$
$$\tag{6.27}$$

Notice that F_s has no dependence on x_s. This method is very similar to that used in [3]. A factor u_s appears for every critical edge of type s appearing in the subtree. The first terms $Z_s - R_s$ give the probability that zero critical nodes are encountered. The second term, with factor u_s, counts the cases where the first node encountered is a critical one. This node may have outgoing edges leading to further critical nodes. These edges are counted by the function F_s, and the use of the generating functions H_n as arguments recursively counts the size of the critical subtree reached upon following each of these edges.

The mean size of the avalanche caused by the removal of single vertex is then given by

$$\sum_s \partial_{u_s} H_s(\mathbf{1}). \tag{6.28}$$

Where ∂_z signifies the partial derivative with respect to variable z.

Let us first examine the mean avalanche size in the case of two layers. Taking partial derivatives of Eqs. (6.26) and (6.27), and after some rearranging, we arrive at

$$\partial_{u_1} H_1(1, 1) = \frac{R_1}{1 - \partial_{x_2} F_1(Z_1, Z_2) \partial_{x_1} F_2(Z_1, Z_2)}. \tag{6.29}$$

where we have used that $F_1(Z_1, Z_2) = R_1$ and also that $H_1(1, 1) = Z_1$, and $H_2(1, 1) = Z_2$.

Let us define the right-hand side of Eq. (6.17) to be $\Psi_1(Z_1, Z_2)$. From Eq. (6.17), and comparing with Eq. (6.27), the partial derivatives of $\Psi_1(Z_1, Z_2)$, are

$$\frac{\partial \Psi_1}{\partial Z_1} = 0$$

$$\frac{\partial \Psi_1}{\partial Z_2} = p(1-f) \sum_{q_1, q_2} \frac{P_{q_1, q_2}}{\langle q_1 \rangle} q_1 q_2 (1-Z_2)^{q_2-1}$$

$$= \frac{\langle q_2 \rangle}{\langle q_1 \rangle} \frac{\partial}{\partial x_1} F_2(Z_1, Z_2). \tag{6.30}$$

and similarly for $\partial\Psi_2/\partial Z_1$ and $\partial\Psi_2/\partial Z_2$. Substituting back, we find that

$$\partial_u H_1(1,1) = \frac{R_1}{(\partial\Psi_1/\partial Z_2)(\partial\Psi_2/\partial Z_1)}. \tag{6.31}$$

The denominator remains finite, and the numerator does not diverge, so this quantity remains finite everywhere in the 2-layer WPP model. This confirms that a discontinuous transition does not occur when there are only two layers.

Following the same procedure in the case of three layers reveals that

$$\partial_{u_1} H_1(1,1) = R_1 \left\{ 1 - \frac{\partial_2\Psi_1[\partial_1\Psi_2 + \partial_1\Psi_3\partial_3\Psi_2]}{1 - \partial_2\Psi_3\partial_3\Psi_2} - \partial_1\Psi_3\partial_3\Psi_1 \right\}^{-1} = \frac{R_1}{1 - \frac{d\Psi_1}{dZ_1}}, \tag{6.32}$$

where for compactness we have written $\partial_m\Psi_n$ for $\partial\Psi_n/\partial Z_m$. Now, an alternative form for the condition for the location of the discontinuous transition is $\frac{d\Psi_1}{dZ_1} = 1$. We see immediately that this implies that the mean avalanche size diverges at the critical point. In other words the avalanches diverge in size as the discontinuous hybrid transition approaches, just as the susceptibility does for an ordinary second-order transition.

In the case of the activation process, WBP, a critical vertex is one that fails the activation criterion for a single type of edge. That is, it has exactly zero edges leading to the root of type s subtrees (probability Z_s), and at least one of every other type. If such a node gains a single connection to the root of a type s subtree, it will itself become the root of such a subtree. Chains of such connections therefore delineate the paths of avalanches of spreading activation. An example of a small critical cluster is shown in Fig. 6.15.

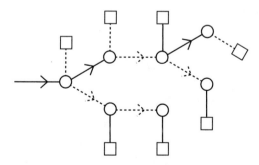

Fig. 6.15 An example of a critical cluster in WBP. Avalanches of activation propagate through the cluster following the *arrowed edges*. If an upstream vertex is activated, all downstream critical vertices will in turn be activated. Note that, unlike for WPP, in WBP it is not possible for an edge to be arrowed in both directions. Activation can only ever propagate in one direction along a given edge. Also note that in the WBP case outgoing critical edges must be of the same type as the incoming one (After [4])

To examine these avalanches, we now define R_s to be the probability that, on following an edge of type s, we encounter a vertex which is not a seed vertex (probability $1 - f$), has not been removed due to random damage (probability p) has at least one child edge of all other types $n \neq s$ leading to the appropriate subtrees (probability Z_n), and zero of type s. That is

$$R_s = p(1 - f) \sum_{q_a, q_b, \dots} \frac{q_s P(q_a, q_b, \dots)}{\langle q_s \rangle} (1 - Z_s)^{q_s - 1} \prod_{n \neq s} [1 - (1 - Z_n)^{q_n}]. \qquad (6.33)$$

Note that this is identical to (6.25), but the probabilities Z_n are different, as is the following argument.

Because critical vertices are outside the WBP cluster, the probabilities Z_s and R_s are mutually exclusive. This means that, upon following an edge of type s, there are three mutually exclusive possibilities: (i) we encounter a subtree of type s (probability Z_m) (ii) we encounter a critical vertex (probability R_s) or (iii) we encounter neither (probability $1 - Z_s - R_s$). We can then define a generating function for the size of the critical subtree encountered upon following an edge of type s (and hence resulting activation avalanche should the parent vertex of that edge be activated) in a recursive way by

$$H_s(\mathbf{u}) = 1 - Z_s - R_s + u_s F_s[H_1(\mathbf{u}), H_2(\mathbf{u}), \dots, H_m(\mathbf{u})]. \qquad (6.34)$$

The functions $F_s(\mathbf{x})$ are defined to be

$$F_s(x, y) = p(1 - f) \sum_{q_a, q_b, \dots} \frac{q_s P(q_a, q_b, \dots)}{\langle q_s \rangle} x_s^{q_s - 1} \prod_{n \neq s} \sum_{l=1}^{q_n} \binom{q_n}{l} Z_n^l x_n^{q_n - l}. \qquad (6.35)$$

Note that $F_s(1 - Z_1, 1 - Z_2, \dots, 1 - Z_m) = R_s$ and $H_s(\mathbf{1}) = 1 - Z_s$.

The mean size of the avalanche caused by the activation of a single vertex is again given by

$$\sum_s \partial_{u_s} H_s(\mathbf{1}). \qquad (6.36)$$

Let us consider the case of WBP in a 2-layer multiplex. Taking partial derivatives of (6.34) and (6.35) and after some rearranging, we find

$$\partial_{u_1} H_1(1, 1) = \frac{R_1 [1 - \partial_{x_2} F_2]}{[1 - \partial_{x_1} F_1][1 - \partial_{x_2} F_2] - \partial_{x_2} F_1 \partial_{x_1} F_2} \qquad (6.37)$$

where for brevity we have not written the arguments of the derivatives of the functions F_1 and F_2, but they should be taken to be evaluated at $(1 - Z_1, 1 - Z_2)$, and where we have used that $F_1(1 - Z_1, 1 - Z_2) = R_1$, $H_1(1, 1) = 1 - Z_1$, and $H_2(1, 1) = 1 - Z_2$.

Remembering that we have defined $\Phi_s(Z_1, Z_2)$, in the two layer case, to be the right-hand side of Eq. (6.21),

$$\frac{\partial \Phi_1}{\partial Z_1} = p(1-f) \sum_{q_1, q_2} \frac{P_{q_1, q_2}}{\langle q_1 \rangle} q_1 (q_1 - 1)(1 - Z_1)^{q_1 - 2} [1 - (1 - Z_2)^{q_2}]$$

$$= \partial_{x_1} F_1 (1 - Z_1, 1 - Z_2) \tag{6.38}$$

and

$$\frac{\partial \Phi_1}{\partial Z_2} = p(1-f) \sum_{q_1, q_2} \frac{P_{q_1, q_2}}{\langle q_1 \rangle} q_1 q_2 (1 - Z_2)^{q_2 - 1} [1 - (1 - Z_1)^{q_1 - 1}]$$

$$= \frac{\langle q_2 \rangle}{\langle q_1 \rangle} \partial_{x_1} F_2 (1 - Z_1, 1 - Z_2) \tag{6.39}$$

and a similar procedure is followed for Φ_2. This means that the equation for $\partial_{u_1} H_1(1, 1)$ can be written

$$\partial_{u_1} H_1(1, 1) = \frac{R_1 [1 - \partial \Phi_2 / \partial Z_2]}{\det[\mathbf{J} - \mathbf{I}]}. \tag{6.40}$$

where the Jacobian matrix $\mathbf{J}$ has elements $J_{ij} = \partial \Phi_i / Z_j$, and $\mathbf{I}$ is the identity matrix. The condition $\frac{d\Phi_1}{dZ_1} = 1$ for the location of the discontinuity in Z_1 (and Z_2) can be rewritten

$$\det[\mathbf{J} - \mathbf{I}] = 0 \tag{6.41}$$

meaning that $\partial_{u_1} H_1(1, 1)$ diverges, and hence the mean avalanche size, diverges precisely at the critical point. This indicates that indeed a discontinuous hybrid transition, with accompanying avalanches of activations, appears even in the two-layer multiplex. A similar analysis can be performed for three or more layers.

6.4 Conclusions

In conclusion, the study of percolation in multiplex networks requires new definitions of connectivity. We have studied the robustness of multiplex networks to damage under two different definitions of connectedness. In the first definition, that is a natural generalization of the concept of single network connectedness, we find a strong criterion which leads to an abrupt collapse of the giant component of a multiplex network having two or more layers. In contrast to ordinary networks, where two vertices are connected if there is a path between them, in multiplex network with m types of edges, two vertices are m-connected if for every kind of edge there is a path from one to another vertex. The transition is a discontinuous hybrid transition, similar to what was found, for example, in the network k-core problem. The collapse occurs through avalanches which diverge in size when the transition is approached from above. We described critical clusters associated with these avalanches. The avalanches are responsible for both the critical scaling and the discontinuity observed in the size of the giant viable cluster.

We compared this with a weaker definition of connectedness, but one which can be calculated locally. In this definition, nodes are members of a cluster if they have at least one edge of each type leading to another member of the cluster. This means that two nodes can belong to the same cluster even when there are no paths of a single color connecting them. We also introduced the concept of invulnerable nodes. In the pruning process form of this model, we find that a two-layer multiplex network no longer exhibits a hybrid transition in the collapse of the giant component, but in three layers such a transition can occur. Finally, we introduced an activation process on multiplex networks, dual to the weak pruning process, in which a small number of seed (invulnerable) nodes are initially activated and further nodes activate if they have connections by every type of edge to active neighbors. The two processes have related phase diagrams, but we find that a discontinuous hybrid transition can occur even when there are only two layers.

Acknowledgements This work was partially supported by the FET IP Project MULTIPLEX 317532 and by the FCT projects EXPL/FIS-NAN/1275/2013 and PEst-C/CTM/LA0025/2011, and post-doctoral fellowship SFRH/BPD/74040/2010, Science Foundation Ireland, Grant No. 11/PI/1026 and the FET-Proactive project PLEXMATH (FP7-ICT-2011-8; Grant No. 317614).

References

1. Baxter, G.J., Dorogovtsev, S.N., Goltsev, A.V., Mendes, J.F.F.: Bootstrap percolation on complex networks. Phys. Rev. E **82**, 011103 (2010)
2. Baxter, G.J., Dorogovtsev, S.N., Goltsev, A.V., Mendes, J.F.F.: Heterogeneous k-core versus bootstrap percolation on complex networks. Phys. Rev. E **83**(5), 051134 (2011)
3. Baxter, G.J., Dorogovtsev, S.N., Goltsev, A.V., Mendes, J.F.F.: Avalanche collapse of interdependent networks. Phys. Rev. Lett. **109**, 248701 (2012)
4. Baxter, G.J., Dorogovtsev, S.N., Mendes, J.F.F., Cellai, D.: Weak percolation on multiplex networks. Phys. Rev. E **89**, 042801 (2014)
5. Baxter, G.J., Dorogovtsev, S.N., Goltsev, A.V., Mendes, J.F.F.: Avalanches in multiplex and interdependent networks. In: D'Agostino, G., Scala, A. (eds.) Networks of Networks: The Last Frontier of Complexity. Understanding Complex Systems, pp. 37–52. Springer International Publishing, Cham (2014)
6. Boccaletti, S., Bianconi, G., Criado, R., del Genio, C.I., Gómez-Gardeñes, J., Romance, M., Sendiña-Nadal, I., Wang, Z., Zanin, M.: The structure and dynamics of multilayer networks. Phys. Rep. **544**, 1–122 (2014)
7. Buldyrev, S.V., Parshani, R., Paul, G., Stanley, H.E., Havlin, S.: Catastrophic casdade of failures in interdependent networks. Nature **464**, 08932 (2010)
8. Caccioli, F., Shrestha, M., Moore, C., Doyne Farmer, J.: Stability analysis of financial contagion due to overlapping portfolios (2012). arXiv:1210.5987

9. Chalupa, J., Leath, P.L., Reich, G.R.: Bootstrap percolation on a bethe lattice. J. Phys. C **12**, L31 (1979)
10. Cohen, R., ben Avraham, D., Havlin, S.: Percolation critical exponents in scale-free networks. Phys. Rev. E **66**, 036113 (2002)
11. Dorogovtsev, S.N., Goltsev, A.V., Mendes, J.F.F.: *k*-core organisation of complex networks. Phys. Rev. Lett. **96**, 040601 (2006)
12. Dorogovtsev, S.N., Goltsev, A.V., Mendes, J.F.F.: Critical phenomena in complex networks. Rev. Mod. Phys. **80**, 1275–1335 (2008)
13. Dueñas, L., Cragin, J.I., Goodno, B.J.: Seismic response of critical interdependent networks. Earthq. Eng. Struct. Dyn. **36**, 285–306 (2007)
14. Gao, J., Buldyrev, S.V., Havlin, S., Stanley, H.E.: Robustness of a network of networks. Phys. Rev. Lett. **107**, 195701 (2011)
15. Huang, X., Vodenska, I., Havlin, S., Stanley, H.E.: Cascading failures in bi-partite graphs: model for systemic risk propagation. Sci. Rep. **3**, 1219 (2013)
16. Leicht, E.A., D'Souza, R.M.: Percolation on interacting networks (2009). arXiv:0907.0894
17. Min, B., K.-I. Goh.: Multiple resource demands and viability in multiplex networks. Phys. Rev. E **89**, 040802(R) (2014)
18. Newman, M.E.J., Strogatz, S.H., Watts, D.J.: Random graphs with arbitrary degree distributions and their applications. Phys. Rev. E **64**, 026118 (2001)
19. Pocock, M.J.O., Evans, D.M., Memmott, J.: The robustness and restoration of a network of ecological networks. Science **335**, 973–976 (2012)
20. Poljanšek, K., Bono, F., Gutiérrez, E.: Seismic risk assessment of interdependent critical infrastructure systems: the case of European gas and electricity networks. Earthq. Eng. Struct. Dyn. **41**, 61–79 (2012)
21. Rinaldi, S.M., Peerenboom, J.P., Kelly, T.K.: Identifying, understanding, and analyzing critical infrastructure interdependencies. IEEE Control Syst. Mag. **21**, 11–25 (2001)
22. Son, S.-W., Bizhani, G., Christensen, C., Grassberger, P., Paczuski, M.: Percolation theory on interdependent networks based on epidemic spreading. EPL **97**, 16006 (2012)

Chapter 7
How Much Interconnected Should Networks be for Cooperation to Thrive?

Zhen Wang, Attila Szolnoki, and Matjaž Perc

Abstract While the consensus is that interconnectivity between networks does promote cooperation by means of organizational complexity and enhanced reciprocity that is out of reach on isolated networks, we here address the question just how much interconnectivity there should be. The more the better according to naive intuition, yet we show that in fact only an intermediate density of sufficiently strong interactions between networks is optimal for the evolution of cooperation. This is due to an intricate interplay between the heterogeneity that causes an asymmetric strategy flow because of the additional links between the networks, and the independent formation of cooperative patterns on each individual network. Presented results are robust to variations of the strategy updating rule, the topology of interconnected networks, and the governing social dilemma, and thus indicate a high degree of universality. We also outline future directions for research based on coevolutionary games and survey existing work.

7.1 Introduction

Network reciprocity is amongst the most well-known mechanisms that may sustain cooperation in evolutionary games that constitute a social dilemma [1]. It was discovered by Nowak and May [2], who observed that on structured populations cooperators can aggregate into compact clusters and so avoid being wiped out

Z. Wang
Department of Physics, Hong Kong Baptist University, Kowloon Tong, Hong Kong

Center for Nonlinear Studies and the Beijing-Hong Kong-Singapore Joint Center for Nonlinear and Complex Systems, Hong Kong Baptist University, Kowloon Tong, Hong Kong

A. Szolnoki
Institute of Technical Physics and Materials Science, Research Centre for Natural Sciences, Hungarian Academy of Sciences, P.O. Box 49, H-1525 Budapest, Hungary

M. Perc (✉)
Faculty of Natural Sciences and Mathematics, University of Maribor, Koroška cesta 160, SI-2000 Maribor, Slovenia
e-mail: matjaz.perc@uni-mb.si

© Springer International Publishing Switzerland 2016   125
A. Garas (ed.), *Interconnected Networks*, Understanding Complex Systems,
DOI 10.1007/978-3-319-23947-7_7

by defectors. Although the mechanism may not work equally well for all social dilemmas [3], and recent empirical evidence based on large-scale economic experiments indicate that it may be compromised or fail altogether [4, 5], there is still ample interest in understanding how and why networks influence the evolution of cooperation. Recent reviews are a testament to the continued liveliness of this field of research [6–9].

Following the explorations of evolutionary games on individual small-world [10–14], scale-free [15–29], coevolving [30–35], hierarchical [36] and bipartite [37] networks, the attention has recently been shifting towards interconnected networks [38–42]. The latter have been put into the spotlight by Buldyrev et al. [43], showing that even seemingly irrelevant changes in one network can have catastrophic and very much unexpected consequence in another network. Subsequently, interconnected networks have been tested for their robustness against attack and assortativity [44–47], properties of percolation [48–52] and diffusion [53], and they have indeed become a hot topic of general interest [54, 55], touching upon subjects as diverse as epidemic spreading [56], the appearance and promotion of creativity [57], and voting [58].

Previous research concerning evolutionary games on interconnected networks has revealed, for example, that biased utility functions suppress the feedback of individual success, which leads to a spontaneous separation of characteristic time scales on the two interconnected networks [38]. Consequently, cooperation is promoted because the aggressive invasion of defectors is more sensitive to the deceleration. Even if the utilities are not biased, cooperation can still be promoted by means of interconnected network reciprocity [39], which however requires simultaneous formation of correlated cooperative clusters on both networks. It has also been shown that the coupling of the evolutionary dynamics in each of the two networks enhances the resilience of cooperation, and that this is intrinsically related to the non-trivial organization of cooperators across the interconnected layers [40]. Perhaps most closely related to the setup of the present work is that by Wang et al. [42], who showed that probabilistic interconnections between interconnected networks can very much promote the evolution of cooperation. In our model, however, the strategy transfer between networks is prohibited. The interconnectivity is thus due solely to coupling together the payoffs of select players that reside on different networks.

Here, we wish to determine how strong the interconnectivity between the networks really ought to be for the optimal promotion of cooperation. Since existing works unequivocally declare that interconnectivity works in favor of the resolution of social dilemmas, one might intuitively assume that the stronger the interconnectivity the better. As we will show, however, this assumption is not necessarily true. To address the problem, we consider primarily the prisoner's dilemma game on two square lattices, where a certain fraction of randomly selected players is allowed to connect with the corresponding players in the other lattice (see Methods section). While strategy transfers between the two networks are not allowed, the additional connections between the corresponding players do influence their utility, and thus their ability to retain and possibly spread their strategies on the home network. This introduces two new parameters, namely the fraction of players that is allowed to form links with the corresponding players in the other network ρ, and the strength of this links α. Together, these two parameters determine the strength of interconnectivity, and also the success of resolving social dilemmas. In the next section we provide further details with regards to the studied evolutionary games, the applied dynamical rule, and the topology of interaction networks. Independent of the strategy updating rule, the topology of interconnected networks, and the governing social dilemma, we will show that cooperation is promoted best if only an intermediate fraction of players is allowed to have external links to the other network, but also that those links should be sufficiently strong. We will also reveal mechanisms that lead to the emergence of the optimal interconnectivity.

7.2   Methods

The evolutionary game is staged on two square lattices, each of size $L \times L$, where initially each player x is designated either as a cooperator ($s_x = C$) or defector ($s_x = D$) with equal probability. Likewise randomly, a fraction ρ of players on each lattice is selected and allowed to form an external link with a corresponding player in the other lattice. Although we predominantly use the square lattice, we will also resort to using the triangle lattice, given that the difference in the clustering coefficient has been determined as a potentially key factor for the outcome of games that are governed by pairwise interactions [67, 68].

The accumulation of payoffs π_x on both networks follows the same standard procedure, depending on the type of the governing social dilemma. In both games two cooperators facing one another acquire R, two defectors get P, whereas a cooperator receives S if facing a defector who then gains T. The prisoner's dilemma game is characterized by the temptation to defect $T = b$, reward for mutual cooperation $R = 1$, and punishment P as well as the sucker's payoff S equaling 0, whereby $1 < b \le 2$ ensures a proper payoff ranking [2]. We note that qualitatively similar results are obtained also for other values of S. The snowdrift game, on the other hand, has $T = \beta$, $R = \beta - 1/2$, $S = \beta - 1$ and $P = 0$, where the temptation to defect can be expressed in terms of the cost-to-benefit ratio $r = 1/(2\beta - 1)$ with $0 \le r \le 1$. Due to the interconnectivity (external links between corresponding players), however, the utilities used to determine fitness are not simply payoffs obtained from the interactions with the nearest neighbors on each individual network, but rather $U_x = \pi_x + \alpha\pi_{x'}$ for players that have an external link, and $U_x = \pi_x$ otherwise. The parameter $0 \le \alpha \le 1$ determines the strength of external links, i.e., the larger its value the higher the potential increase of utility of two players that are connected by the external link.

Importantly, while the interconnectivity affects the utility of players, it does not allow strategies to be transferred between the two networks. Thus, the evolution of the two strategies proceeds in accordance with the standard Monte Carlo simulation procedure comprising the following elementary steps for each network. First, a randomly selected player x acquires its utility U_x by playing the game with all its nearest neighbors and taking into account also the potential addition to the utility stemming from the possible external link, as described above. Next, one randomly chosen neighbor of x within the same network, denoted by y, also acquires its utility U_y in the same way. Lastly, player x attempts to adopt the strategy s_y from player y with a probability determined by the Fermi function

$$W(s_y \rightarrow s_x) = \frac{1}{1 + \exp[(U_x - U_y)/K]} , \qquad (7.1)$$

where $K = 0.1$ quantifies the uncertainty related to the strategy adoption process [6, 65]. The latter is usually associated with errors in decision making and imperfect information transfer between the players. Notably, to test the robustness of our findings, we will also use the best-takes-over [2] and the proportional imitation [66] strategy updating rule. Regardless of which type of interaction network, evolutionary game, or the strategy updating rule is used, in accordance with the random asynchronous update, each player on both networks is selected once on average during a full Monte Carlo step. Moreover, sufficiently large system sizes (from $L = 200$ to $L = 800$) and relaxation times need to be used to avoid finite size effects and to ensure a stationary state has been reached. Presented results were averaged over up to 30 independent runs to further improve accuracy.

7.3   Results

To begin with, we show in Fig. 7.1 the impact of parameters ρ and α on the outcome of the prisoner's dilemma game. It can be observed that there exists an intermediate range of the fraction of players that are allowed to form an external link at which

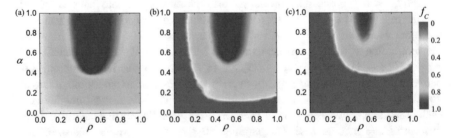

Fig. 7.1 Tuning in on the optimal interconnectivity between two square lattices for the resolution of the prisoner's dilemma. Color coded is the fraction of cooperators f_C in dependence on the fraction of players that are allowed to form an external link ρ and the strength of these links α, as obtained for $b = 1.03$ (**a**), $b = 1.05$ (**b**) and $b = 1.1$ (**c**). Irrespective of b, there exists an intermediate value of $\rho \approx 0.5$ at which cooperation is optimally promoted. But in addition to that, the value of α needs to be sufficiently large as well (After [70])

cooperators fare best. Irrespective of the temptation to defect b, values around $\rho \approx 0.5$ yield an optimal outcome of the social dilemma. Yet the coupling strength is important too. Only if the value of α is sufficiently large are the players able to utilize the advantage of being linked to their corresponding players in the other network. Although the level of cooperation appears to fade slightly beyond $\alpha = 0.7$ if the temptation to defect is low or moderate [panels (a) and (b)], the prevailing conclusion is that the coupling strength needs to be sufficiently strong.

To clarify the mechanism that is responsible for the promotion of cooperation, we first monitor the evolution of cooperation by measuring not just the overall average cooperation level (f_C), but separately also the average cooperation level for players with (f_{C_d}) and without (f_{C_o}) external links to the other network in dependence on the number of full Monte Carlo steps (MCS), as defined in the Methods section. For easier reference, we will refer to individuals with external links to the other network as "distinguished" and to those who have no such links as "ordinary" players (note that the subscripts in f_{C_d} and f_{C_o} are chosen accordingly). Figure 7.2 reveals that the cooperation level amongst the distinguished players who do have external links to the other network is significantly higher than the cooperation level amongst the players who are not externally linked. Expectedly, the overall cooperation level is in-between f_{C_d} and f_{C_o}. The identified difference between f_{C_d} and f_{C_o} is crucial, because it indicates that players who have the ability to collect an additional payoff from the other network are more likely to cooperate. Indeed, the natural selection

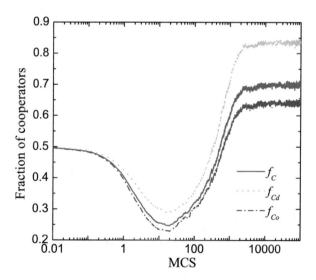

Fig. 7.2 Distinguished players who have an external link with their corresponding player in the other network are more likely to cooperate than those who are not externally linked. Depicted is the time evolution of the fraction of cooperators in the whole population (f_C), among the distinguished players (f_{C_d}), and among ordinary players, i.e., those that do not have an external link to the other network (f_{C_o}). It can be observed that $f_{C_d} > f_{C_o}$. Parameter values used were: $b = 1.05$, $\rho = 0.3$ and $\alpha = 0.8$ (After [70])

of the cooperative strategy among distinguished players is higher than in the whole population, which is a consequence of an asymmetric strategy flow that emerges between the distinguished and other players. Because of generally higher payoffs, the distinguished players are followed by the others, which results in the selection of cooperation around them. In other words, distinguished players with an external link to the other network play the role of leaders in the community, similarly as was reported many times before for hubs on heterogeneous isolated (individual) networks [6].

To test this explanation further, we directly adjust the teaching activity of players, i.e., the ability to pass their strategy to a neighbor [59, 60], which can be done effectively by introducing a multiplication factor w to Eq. (7.1). We consider two options. First, we depreciate all distinguished players by using $w = 0.05$ while keeping $w = 1$ for those who do not have an external link, and second, we reverse these values. The expectation is that the first option will nullify the advantage of interconnectivity between the two networks, while the second option will further amplify the positive effects on the evolution of cooperation.

Figure 7.3 reveals nicely how the leading role of distinguished players improves the cooperation level [panels (a) and (c)], yet only if distinguished players are not depreciated by $w = 0.05$ [panel (b)]. For comparison, we show in panel (a) the results obtained with the basic model where all players have $w = 1$. There the change from $\alpha = 0$ to $\alpha = 0.5$ (note that for $\alpha = 0$ the difference between distinguished and ordinary players vanishes) introduces a noticeable increase in the critical temptation to defect where cooperators die out and an overall increase in the level of cooperation. But if $w = 0.05$ is applied to distinguished players,

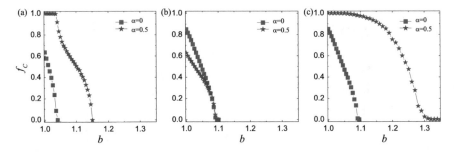

Fig. 7.3 Interconnectivity between networks favors cooperation only if the awarded additional payoffs are not counteracted by reduced teaching activity. Panel (**a**) depicts the fraction of cooperators f_C in dependence on the temptation to defect b for the basic version of the game, where all players have teaching activity $w = 1$. In panel (**b**) the teaching activity of all distinguished players who have an external link to the other network is reduced to $w = 0.05$. It can be observed that the promotion of cooperation due to the interconnectivity vanishes. In panel (**c**) the teaching activity of all ordinary players is reduced to $w = 0.05$, which further strengthens the leading role of the distinguished players and leads to the strongest promotion of cooperation. All panels feature results for $\alpha = 0$ and $\alpha = 0.5$ at $\rho = 0.5$, where $\alpha = 0$ means that effectively the two networks are isolated, i.e., there is no interconnectivity because the links between the two networks yield no additional utility to either player (After [70])

then they become unable to lead the others despite of the fact that their utilities are higher than those of their neighbors. In the absence of hierarchy that was previously warranted by the interconnectivity between networks, there will form no groups with homogeneous strategies, and hence the advantage of cooperation will not be revealed. As panel (b) shows quite interestingly, $\alpha = 0$ (absence of interconnectivity between networks) even slightly outperforms $\alpha = 0.5$, because a weak hierarchy is then restored due to the unequal teaching activity of players. Note that at $\alpha = 0.5$ the unequal teaching activity is counteracted by the additional payoff distinguished players receive from the other network. At $\alpha = 0$ this counterbalance is lost, and the ordinary players become the leaders due to their higher value of w. The positive effect on the level of cooperation is, however, rather marginal.

In the opposite case, when distinguished players are endowed with the full teaching activity $w = 1$ while ordinary players without an external link are depreciated with $w = 0.05$, the change from $\alpha = 0$ to $\alpha = 0.5$ is the strongest [panel (c)]. Here the inequality of w and the additional payoffs stemming from the other network are able to strengthen each other and fortify the leaders to yield the highest cooperation level within the framework of this model. The enforced role of distinguished players will cause an immediate reaction from the followers, and this prompt reaction will select the more successful cooperative strategy as described before. From the technical point of view, it is interesting to note that the curves for $\alpha = 0$ in Fig. 7.3b, c coincide, because in the absence of extra payoff from the other network it does not matter whether distinguished or ordinary players are endowed with $w = 1$ (or $w = 0.05$). Since the fraction of distinguished players was set to $\rho = 0.5$, both fractions are equally large, thus resulting in the same cooperation level.

Results presented in Fig. 7.3 and the pertaining interpretation can be corroborated by comparing the evolution of cooperation within different groups of players, similarly as done in Fig. 7.2 above. As Fig. 7.4b shows, when the teaching activity is reduced for distinguished players the cooperation among them is not favored by natural selection. Consequently, only a very low value of b can ensure a reasonably high cooperation level in the whole population because the latter is dragged down considerably by the low f_{C_d}. There is a slight improvement among other players, as evidenced by the higher f_{C_o}, which is due to a higher teaching activity. However, despite their higher teaching activity, ordinary players cannot lead the whole population efficiently because distinguished players are reluctant to follow them due to their higher individual utility. If either the teaching activity of all players is left intact [panel (a)] or the higher teaching activity is awarded to distinguished players [panel (c)], then $f_{C_o} < f_{C_d}$, as observed initially in Fig. 7.2. Thus, the basic mechanism of cooperation promotion is restored or even additionally fortified.

Furthermore, it is important to emphasize that the optimal interconnectivity between the networks can work in favor of cooperation only if there are distinguished players in both graphs. On the other hand, if the additional utilities flow only in one direction, the mechanism will fail or yield only a marginally better outcome. Results supporting this argumentation are presented in Fig. 7.5. For reference, the

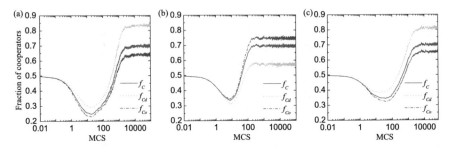

Fig. 7.4 Decreasing the teaching activity of distinguished players nullifies their higher propensity to cooperate, and it reduces the temptation to defect that still warrants a healthy cooperation level in the whole population. Depicted is the time evolution of the fraction of cooperators in the whole population (f_C), among the distinguished players (f_{C_d}), and among ordinary players (f_{C_o}). It can be observed that $f_{C_d} > f_{C_o}$ in panels (**a**) and (**c**), but not in panel (**b**). In the latter, the teaching activity of all distinguished players who have an external link to the other network is reduced to $w = 0.05$. In panel (**a**) all players have $w = 1$, while in panel (**c**) the teaching activity of all ordinary players is reduced to $w = 0.05$, which further amplifies the $f_{C_d} > f_{C_o}$ difference. We have used parameter values that yield approximately the same overall level of cooperation (f_C) in all three cases: $\rho = 0.3$, $b = 1.05$ and $\alpha = 0.8$ for panel (**a**), $\rho = 0.3$, $b = 1.0$ and $\alpha = 0.25$ for panel (**b**), and $\rho = 0.3$, $b = 1.09$ and $\alpha = 0.25$ for panel (**c**) (After [70])

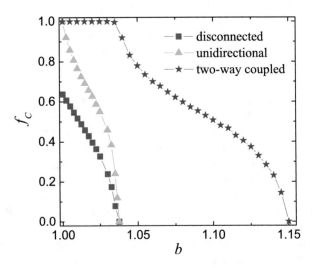

Fig. 7.5 Optimal interconnectivity between the two networks works optimally only if the links connecting them go both ways. Depicted is the fraction of cooperators f_C in dependence on the temptation to defect b as obtained on an isolated network (reference case), on two networks connected by means of unidirectional links, and on two mutually interconnected networks. Cooperation is optimally promoted only if there is independent formation of cooperative patterns on each individual network, for which the chance for heterogeneity (distinguished players) needs to be provided on both of them. Parameter values used were: $\rho = 0.5$ and $\alpha = 0.5$ (After [70])

outcome on an isolated network is depicted as well. It can be observed that if distinguished players in the first network can collect additional payoffs from the second network, but the evolution in the second network is independent from the first network because players there are unable to collect additional payoffs (the interconnectivity is unilateral), then the critical temptation to defect b at which cooperators die out does not increase at all. In fact, only the level of cooperation increases slightly in the mixed $C + D$ phase. If the interconnectivity is bilateral, i.e., players on both networks can collect an additional payoff from the other network, however, the positive impact on the evolution of cooperation is much stronger. This highlights the importance of mutual interconnectivity and the independent formation of cooperative patterns on each individual network. Players connected to their corresponding partners in the other network can support each other effectively only if homogeneous cooperative domains emerge on both networks. Notably, the importance of correlated growth and formation of cooperative domains has been emphasized already in [39], although here they need not overlap geographically. We could observe practically the same cooperation level when distinguished players collected additional payoffs from an ordinary player in the other network. Namely, there is no need to link distinguished players from different networks with one another. The crucial condition is the chance of heterogeneity on both networks [61], the positive effect of which can then be mutually amplified through the interconnectivity.

Our argument for the spreading of cooperative behavior among distinguished players can also be supported by how the border of the full D phase behaves in dependence of ρ and α at a high temptation to defect. As results presented in Fig. 7.1c suggest, a smaller density of distinguished players can be compensated by a higher value of α, but only up to a certain point. At such a high temptation value cooperators cannot survive even for large values of α if the density of distinguished players ρ is below a threshold value. Naturally, the critical ρ depends slightly on the temptation to defect [compare panels (b) and (c)], but the smallest value of the phase transition point is close to the critical $\rho_c = 0.1869(1)$ value of jamming coverage of particles during a random sequential adsorption when nearest and next nearest neighbor interactions are excluded on a square lattice [62]. Shortly, if distinguished players are too rare, then to support them via a high α will not yield the desired impact because their influence cannot percolate. The latter, however, is an essential condition to maintain cooperation, which was already pointed out in previous works [63, 64].

It is lastly of interest to verify the robustness of these observations, first with regards to the strategy updating rule. As results presented in Fig. 7.6 demonstrate, our conclusions are not restricted solely to the Fermi-type strategy updating [6, 65], but remain valid also under the best-takes-over rule [2] and proportional imitation [66]. In both cases an optimal intermediate value of ρ is clearly inferable, and the positive effect on the evolution of cooperation is the stronger the larger the value of the coupling strength α. This is qualitatively identical as observed in Fig. 7.1 with the Fermi rule.

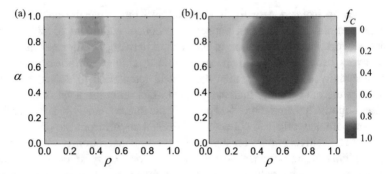

Fig. 7.6 Robustness of optimal network interconnectivity on different strategy updating rules. Color coded is the fraction of cooperators f_C in dependence on the fraction of players that are allowed to form an external link ρ and the strength of these links α, as obtained for $b = 1.4$ and best-takes-over strategy updating (**a**), and $b = 1.06$ and proportional imitation (**b**). Irrespective of the applied strategy updating rule, there exists an intermediate value of ρ at which cooperation is optimally promoted, and the dependence on α is also qualitatively the same as in Fig. 7.1, where the Fermi strategy adoption rule has been used (After [70])

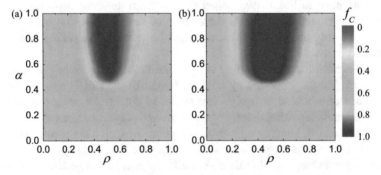

Fig. 7.7 Robustness of optimal network interconnectivity on the topology and the type of social dilemma. Color coded is the fraction of cooperators f_C in dependence on the fraction of players that are allowed to form an external link ρ and the strength of these links α, as obtained for $b = 1.1$ on the triangle lattice (**a**), and $r = 0.3$ (snowdrift game) on the square lattice (**b**). The Fermi strategy adoption rule has been used in both cases. Irrespective of the topology of each individual network and the type of the governing social dilemma, the results are qualitatively the same as in Figs. 7.1 and 7.6 (After [70])

Since previous works revealed that the clustering coefficient could be a decisive factor affecting the evolution of cooperation in games that are governed by pairwise interactions [67, 68] (note that this is not the case for games governed by group interactions [69]), it is also instructive to examine the relevance of network interconnectivity under this condition. Unlike the square lattice that has a zero clustering coefficient, the triangular lattice has a high clustering coefficient, and thus serves the purpose very well. Results presented in Fig. 7.7a attest to the fact that the existence of optimal network interconnectivity does not depend on structural properties of each individual network, as indeed the $\alpha - \rho$ dependence of f_C is the same as observed before for the square lattice in Fig. 7.1. To conclude, we extend our exploration also to other social dilemmas, more precisely the snowdrift game, and as shown in Fig. 7.7b, the main conclusions remain intact. To reiterate, the exists and intermediate level of interconnectivity between networks that is optimal for the resolution of social dilemmas.

7.4 Discussion

Summarizing, we have studied the evolution of cooperation in the spatial prisoner's dilemma and the snowdrift game subject to interconnectedness by means of a different fraction of differently strong links between the corresponding players residing on the two interconnected networks. We have found that for cooperation to be optimally promoted, the interconnectivity should stem only from an intermediate fraction of links connecting the two networks, and that those links should affect the utility of players significantly. The existence of optimal interconnectivity has been attributed to the heterogeneity that is brought about by the enhanced utility of those players that do have external links to the other network, as opposed to those who have not. This introduces asymmetric strategy flow, which in turn leads to the emergence of influential leaders that can act as strong cooperative hubs in their respective networks. Importantly, the compact cooperative patterns that appear independently on both networks support each other mutually through the links that constitute the interconnectivity. Indeed, we have shown that the mechanism works best only if the interconnectivity is bilateral, and if the asymmetric strategy flow is not counteracted by artificial low weights assigned to the reproducibility of interconnected players. In case of unidirectional interconnectivity or if the reproducibility of players is altered, however, some marginal benefits for cooperators may still exist, but these are then far removed from the full potential of interconnected networks to aid the resolution of social dilemmas. We have tested the robustness of these conclusions by replacing the Fermi strategy adoption rule with the best-takes-over and the proportional imitation rule, as well as by replacing the square lattice having zero clustering coefficient with the triangle lattice that has a much higher clustering coefficient, as well as finally by replacing the prisoner's dilemma game with the snowdrift game. Quite remarkably, we have found that the optimal interconnectivity persist across all these different setups, thus leading to the conclusion that it ought to be to a large degree a universally valid phenomenon.

While interconnectivity can promote the evolution of cooperation by means of the mechanisms presented in this chapter (for further details see [38–41]), and while it has been established that there in fact exists an optimal intermediate level of interconnectivity that works best in deterring defection [70, 71], it is not clear how it could have emerged from initially isolated subsystems. Many works have simply assumed some level of interconnectivity be there without considering its origin or mechanisms that might have led to its emergence. To amend this, we have recently adopted the established concept of coevolution in the realm of the prisoner's dilemma game [8], with the aim of introducing an elementary coevolutionary rule that leads towards the spontaneous emergence of optimal interconnectivity between two initially completely independent networks [72]. The rule is simple indeed, assuming only that players who are often enough successful at passing their strategy to one of their neighbors, regardless of which strategy it is, are allowed to form an external link to the corresponding player in the other network in order to potentially increase their utility. As we have shown [72], this suffices for the system to self-organize into a two-class society in which the resolution of the prisoner's dilemma is most likely. Alternatively, it is possible to introduce rewards for high-enough evolutionary fitness of individual players in the form of additional links that bridge the gap between two initially disconnected populations [73]. These rewards effectively introduce interconnectivity between two populations, and they allow the rewarded players to increase their utility with a fraction of the utility of the player in the other population. Regardless of game-specific details, the self-organization of fitness and reward promotes the evolution of cooperation well past the boundaries imposed by traditional network reciprocity [2], as well as past the boundaries imposed by interconnected network reciprocity [39], if only the utility threshold is sufficiently large [73]. On the other hand, the threshold must not exceed a critical value, which could be well below the maximal possible utility a cooperator is able to reach if it would be fully surrounded by other cooperators. These latest results concerning coevolution on interconnected networks indicate that even seemingly irrelevant and minute additions to the basic evolutionary process might have led to the intricate and widespread interconnectivity between networks that we witness today in many social and technological systems.

While the games on interconnected networks studied here are not meant to model a particular real-life situation, they nevertheless do capture the essence of some situations that are viable in reality. For example, it is generally accepted that not all individuals are equally fond of making connections outside of their natural environment. Similarly, some would very much wish to do so, but may not have a chance. These and similar considerations may all affect the level of interconnectivity between two or more networks, and it is within the realm of these possibilities that our study predicts the existence of an optimal level of interconnectedness. Future studies could address the coupling of more complex (small world or scale-free, for example) interaction topologies, the outcome of other games on interconnected networks, such as for example the traveler's dilemma game that has recently been studied in a spatial setting [74], as well as the impact of coevolution and growth, both of which have recently been the subject of much interest [75–79]. Overall, it seems safe to conclude that the interconnectivity of interaction networks offers several exciting possibilities for further research related to evolutionary games, and it ought to bring the models a step closer to actual conditions, given that networks indeed rarely exist in an isolated state.

Acknowledgements This research was supported by the Hungarian National Research Fund (Grant K-101490), TAMOP-4.2.2.A-11/1/KONV-2012-0051, and the Slovenian Research Agency (Grants J1-4055 and P5-0027).

References

1. Nowak, M.A.: Five rules for the evolution of cooperation. Science **314**, 1560–1563 (2006)
2. Nowak, M.A., May, R.M.: Evolutionary games and spatial chaos. Nature **359**, 826–829 (1992)
3. Hauert, C., Doebeli, M.: Spatial structure often inhibits the evolution of cooperation in the snowdrift game. Nature **428**, 643–646 (2004)
4. Gracia-Lázaro, C., Cuesta, J., Sánchez, A., Moreno, Y.: Human behavior in prisoner's dilemma experiments suppresses network reciprocity. Sci. Rep. **2**, 325 (2012)
5. Gracia-Lázaro, C., Ferrer, A., Ruiz, G., Tarancón, A., Cuesta, J., Sánchez, A., Moreno, Y.: Heterogeneous networks do not promote cooperation when humans play a prisoner's dilemma. Proc. Natl. Acad. Sci. U. S. A. **109**, 12922–12926 (2012)
6. Szabó, G., Fáth, G.: Evolutionary games on graphs. Phys. Rep. **446**, 97–216 (2007)
7. Roca, C.P., Cuesta, J.A., Sánchez, A.: Evolutionary game theory: temporal and spatial effects beyond replicator dynamics. Phys. Life Rev. **6**, 208–249 (2009)
8. Perc, M., Szolnoki, A.: Coevolutionary games – a mini review. BioSystems **99**, 109–125 (2010)
9. Perc, M., Gómez-Gardeñes, J., Szolnoki, A., Floría, L.M., Moreno, Y.: Evolutionary dynamics of group interactions on structured populations: a review. J. R. Soc. Interface **10**, 20120997 (2013)
10. Abramson, G., Kuperman, M.: Social games in a social network. Phys. Rev. E **63**, 030901(R) (2001)
11. Kim, B.J., Trusina, A., Holme, P., Minnhagen, P., Chung, J.S., Choi, M.Y.: Dynamic instabilities induced by asymmetric influence: prisoner's dilemma game in small-world networks. Phys. Rev. E **66**, 021907 (2002)
12. Masuda, N., Aihara, K.: Spatial prisoner's dilemma optimally played in small-world networks. Phys. Lett. A **313**, 55–61 (2003)
13. Tomassini, M., Luthi, L., Giacobini, M.: Hawks and doves games on small-world networks. Phys. Rev. E **73**, 016132 (2006)
14. Fu, F., Liu, L.-H., Wang, L.: Evolutionary prisoner's dilemma on heterogeneous Newman-Watts small-world network. Eur. Phys. J. B **56**, 367–372 (2007)
15. Santos, F.C., Pacheco, J.M.: Scale-free networks provide a unifying framework for the emergence of cooperation. Phys. Rev. Lett. **95**, 098104 (2005)
16. Santos, F.C., Pacheco, J.M., Lenaerts, T.: Evolutionary dynamics of social dilemmas in structured heterogeneous populations. Proc. Natl. Acad. Sci. USA **103**, 3490–3494 (2006)
17. Gómez-Gardeñes, J., Campillo, M., Floría, L.M., Moreno, Y.: Dynamical organization of cooperation in complex networks. Phys. Rev. Lett. **98**, 108103 (2007)
18. Rong, Z., Li, X., Wang, X.: Roles of mixing patterns in cooperation on a scale-free networked game. Phys. Rev. E **76**, 027101 (2007)
19. Masuda, N.: Participation costs dismiss the advantage of heterogeneous networks in evolution of cooperation. Proc. R. Soc. B **274**, 1815–1821 (2007)
20. Tomassini, M., Luthi, L., Pestelacci, E.: Social dilemmas and cooperation in complex networks. Int. J. Mod. Phys. C **18**, 1173–1185 (2007)
21. Szolnoki, A., Perc, M., Danku, Z.: Towards effective payoffs in the prisoner's dilemma game on scale-free networks. Physica A **387**, 2075–2082 (2008)
22. Assenza, S., Gómez-Gardeñes, J., Latora, V.: Enhancement of cooperation in highly clustered scale-free networks. Phys. Rev. E **78**, 017101 (2008)
23. Santos, F.C., Santos, M.D., Pacheco, J.M.: Social diversity promotes the emergence of cooperation in public goods games. Nature **454**, 213–216 (2008)
24. Peña, J., Volken, H., Pestelacci, E., Tomassini, M.: Conformity hinders the evolution of cooperation on scale-free networks. Phys. Rev. E **80**, 016110 (2009)
25. Poncela, J., Gómez-Gardeñes, J., Moreno, Y.: Cooperation in scale-free networks with limited associative capacities. Phys. Rev. E **83**, 057101 (2011)
26. Brede, M.: Playing against the fittest: a simple strategy that promotes the emergence of cooperation. EPL **94**, 30003 (2011)

27. Tanimoto, J., Brede, M., Yamauchi, A.: Network reciprocity by coexisting learning and teaching strategies. Phys. Rev. E **85**, 032101 (2012)
28. Pinheiro, F., Pacheco, J., Santos, F.: From local to global dilemmas in social networks. PLoS One **7**, e32114 (2012)
29. Simko, G.I., Csermely, P.: Nodes having a major influence to break cooperation define a novel centrality measure: game centrality. PLoS One **8**, e67159 (2013)
30. Ebel, H., Bornholdt, S.: Coevolutionary games on networks. Phys. Rev. E **66**, 056118 (2002)
31. Zimmermann, M.G., Eguíluz, V., Miguel, M.S.: Coevolution of dynamical states and interactions in dynamic networks. Phys. Rev. E **69**, 065102(R) (2004)
32. Pacheco, J.M., Traulsen, A., Nowak, M.A.: Coevolution of strategy and structure in complex networks with dynamical linking. Phys. Rev. Lett. **97**, 258103 (2006)
33. Santos, F.C., Pacheco, J.M., Lenaerts, T.: Cooperation prevails when individuals adjust their social ties. PLoS Comput. Biol. **2**, 1284–1290 (2006)
34. Fu, F., Chen, X., Liu, L., Wang, L.: Promotion of cooperation induced by the interplay between structure and game dynamics. Physica A **383**, 651–659 (2007)
35. Tanimoto, J.: Dilemma solving by coevolution of networks and strategy in a 2 × 2 game. Phys. Rev. E **76**, 021126 (2007)
36. Lee, S., Holme, P., Wu, Z.-X.: Emergent hierarchical structures in multiadaptive games. Phys. Rev. Lett. **106**, 028702 (2011)
37. Gómez-Gardeñes, J., Romance, M., Criado, R., Vilone, D., Sánchez, A.: Evolutionary games defined at the network mesoscale: the public goods game. Chaos **21**, 016113 (2011)
38. Wang, Z., Szolnoki, A., Perc, M.: Evolution of public cooperation on interdependent networks: the impact of biased utility functions. EPL **97**, 48001 (2012)
39. Wang, Z., Szolnoki, A., Perc, M.: Interdependent network reciprocity in evolutionary games. Sci. Rep. **3**, 1183 (2013)
40. Gómez-Gardeñes, J., Reinares, I., Arenas, A., Floría, L.M.: Evolution of cooperation in multiplex networks. Sci. Rep. **2**, 620 (2012)
41. Gómez-Gardeñes, J., Gracia-Lázaro, C., Floría, L.M., Moreno, Y.: Evolutionary dynamics on interdependent populations. Phys. Rev. E **86**, 056113 (2012)
42. Wang, B., Chen, X., Wang, L.: Probabilistic interconnection between interdependent networks promotes cooperation in the public goods game. J. Stat. Mech. **2012**, P11017 (2012)
43. Buldyrev, S.V., Parshani, R., Paul, G., Stanley, H.E., Havlin, S.: Catastrophic cascade of failures in interdependent networks. Nature **464**, 1025–1028 (2010)
44. Huang, X., Gao, J., Buldyrev, S.V., Havlin, S., Stanley, H.E.: Robustness of interdependent networks under targeted attack. Phys. Rev. E **83**, 065101(R) (2011)
45. Zhou, D., Stanley, H.E., D'Agostino, G., Scala, A.: Assortativity decreases the robustness of interdependent networks. Phys. Rev. E **86**, 066103 (2012)
46. Baxter, G.J., Dorogovtsev, S.N., Goltsev, A.V., Mendes, J.F.F.: Avalanche collapse of interdependent networks. Phys. Rev. Lett. **109**, 248701 (2012)
47. Peixoto, T.P., Bornholdt, S.: Evolution of robust network topologies: emergence of central backbones. Phys. Rev. Lett. **109**, 118703 (2012)
48. Parshani, R., Buldyrev, S.V., Havlin, S.: Interdependent networks: reducing the coupling strength leads to a change from a first to second order percolation transition. Phys. Rev. Lett. **105**, 048701 (2010)
49. Son, S.-W., Bizhani, G., Christensen, C., Grassberger, P., Paczuski, M.: Percolation theory on interdependent networks based on epidemic spreading. EPL **97**, 16006 (2012)
50. Lau, H.W., Paczuski, M., Grassberger, P.: Agglomerative percolation on bipartite networks: nonuniversal behavior due to spontaneous symmetry breaking at the percolation threshold. Phys. Rev. E **86**, 011118 (2012)
51. Schneider, C.M., Araújo, N.A.M., Herrmann, H.J.: Algorithm to determine the percolation largest component in interconnected networks. Phys. Rev. E **87**, 043302 (2013)
52. Zhao, K., Bianconi, G.: Percolation on interacting, antagonistic networks. J. Stat. Mech. **2013**, P05005 (2013)

53. Gómez, S., Díaz-Guilera, A., Gómez-Gardeñes, J., Pérez-Vicente, C., Moreno, Y., Arenas, A.: Diffusion dynamics on multiplex networks. Phys. Rev. Lett. **110**, 028701 (2013)
54. Gao, J., Buldyrev, S.V., Stanley, H.E., Havlin, S.: Networks formed from interdependent networks. Nat. Phys. **8**, 40–48 (2012)
55. Havlin, S., Kenett, D.Y., Ben-Jacob, E., Bunde, A., Hermann, H., Kurths, J., Kirkpatrick, S., Solomon, S., Portugali, J.: Challenges of network science: applications to infrastructures, climate, social systems and economics. Eur. J. Phys. Spec. Top. **214**, 273–293 (2012)
56. Wang, Y., Xiao, G.: Epidemics spreading in interconnected complex networks. Phys. Lett. A **376**, 2689–2696 (2012)
57. Csermely, P.: The appearance and promotion of creativity at various levels of interdependent networks. Talent Dev. Excell. **5**, 115–123 (2013)
58. Halu, A., Zhao, K., Baronchelli, A., Bianconi, G.: Connect and win: the role of social networks in political elections. EPL **102**, 16002 (2013)
59. Szolnoki, A., Szabó, G.: Cooperation enhanced by inhomogeneous activity of teaching for evolutionary prisoner's dilemma games. EPL **77**, 30004 (2007)
60. Szolnoki, A., Perc, M.: Coevolution of teaching activity promotes cooperation. New J. Phys. **10**, 043036 (2008)
61. Santos, F.C., Pinheiro, F., Lenaerts, T., Pacheco, J.M.: Role of diversity in the evolution of cooperation. J. Theor. Biol. **299**, 88–96 (2012)
62. Dickman, R., Wang, J.-S., Jensen, I.: Random sequential adsorption: series and virial expansions. J. Chem. Phys. **94**, 8252–8257 (1991)
63. Wang, Z., Szolnoki, A., Perc, M.: Percolation threshold determines the optimal population density for public cooperation. Phys. Rev. E **85**, 037101 (2012)
64. Wang, Z., Szolnoki, A., Perc, M.: If players are sparse social dilemmas are too: importance of percolation for evolution of cooperation. Sci. Rep. **2**, 369 (2012)
65. Blume, L.E.: The statistical mechanics of strategic interactions. Games Econ. Behav. **5**, 387–424 (1993)
66. Schlag, K.H.: Why imitate, and if so, how? A bounded rational approach to multi-armed bandits. J. Econ. Theory **78**, 130–156 (1998)
67. Szabó, G., Vukov, J., Szolnoki, A.: Phase diagrams for an evolutionary prisoner's dilemma game on two-dimensional lattices. Phys. Rev. E **72**, 047107 (2005)
68. Kuperman, M.N., Risau-Gusman, S.: Relationship between clustering coefficient and the success of cooperation in networks. Phys. Rev. E **86**, 016104 (2012)
69. Szolnoki, A., Perc, M., Szabó, G.: Topology-independent impact of noise on cooperation in spatial public goods games. Phys. Rev. E **80**, 056109 (2009)
70. Wang, Z., Szolnoki, A., Perc, M.: Optimal interdependence between networks for the evolution of cooperation. Sci. Rep. **3**, 2470 (2013)
71. Jiang, L.-L., Perc, M.: Spreading of cooperative behaviour across interdependent groups. Sci. Rep. **3**, 2483 (2013)
72. Wang, Z., Szolnoki, A., Perc, M.: Self-organization towards optimally interdependent networks by means of coevolution. New J. Phys. **16**, 033041 (2014)
73. Wang, Z., Szolnoki, A., Perc, M.: Rewarding evolutionary fitness with links between populations promotes cooperation. J. Theor. Biol. **349**, 50–56 (2014)
74. Li, R.-H., Yu, J.X., Lin, J.: Evolution of cooperation in spatial traveler's dilemma game. PLoS One **8**, e58597 (2013)
75. Poncela, J., Gómez-Gardeñes, J., Floría, L.M., Sánchez, A., Moreno, Y.: Complex cooperative networks from evolutionary preferential attachment. PLoS One **3**, e2449 (2008)
76. Poncela, J., Gómez-Gardeñes, J., Traulsen, A., Moreno, Y.: Evolutionary game dynamics in a growing structured population. New J. Phys. **11**, 083031 (2009)
77. Portillo, I.G.: Building cooperative networks. Phys. Rev. E **86**, 051108 (2012)
78. Li, G., Jin, X.-G., Song, Z.-H.: Evolutionary game on a stochastic growth network. Physica A **391**, 6664–6673 (2012)
79. Brede, M.: Short versus long term benefits and the evolution of cooperation in the prisoner's dilemma game. PLoS One **8**, e56016 (2013)

Chapter 8
The Cacophony of Interconnected Networks

Vitor H.P. Louzada, Nuno A.M. Araújo, José S. Andrade Jr., and Hans J. Herrmann

Abstract The harmony of an orchestra emerges from the individual effort of musicians towards mutual synchronization of their tempi. When the orchestra is split between two concert halls communicating via Internet, a time delay is imposed which might hinder synchronization. We present this type of system as two interconnected networks of oscillators with a time delay and analyze its dynamics as a function of the couplings and communication lag. We describe a breathing synchronization regime, namely, for a wide range of parameters, two groups emerge in the orchestra within the same concert hall playing at different tempi. Each group has a mirror in the other hall, one group is in phase and the other in anti-phase with their mirrors. For strong couplings, a phase shift between halls might occur. The implications of our findings on other interconnected systems are also discussed.

Technology has furnished us with global connectivity changing the functioning of cooperative work, international business, and interpersonal relationships. Today, it is possible to distribute an orchestra over two concert halls in different continents. Fast Internet connections would provide the communication infrastructure to properly combine the sounds [12]. As there is always a physical limit speed to information transport, the communication between sub-orchestras is subjected to a time delay due to the distance between halls. As we discuss here, this time delay might pose a real challenge to the synchronizability among musicians, which is vital to a successful performance. When isolated, each sub-orchestra would naturally play

V.H.P. Louzada (✉)
Computational Physics, IfB, ETH Zürich, Wolfgang-Pauli-Strasse 27, 8093 Zürich, Switzerland
e-mail: louzada@ethz.ch

N.A.M. Araújo
Departamento de Física, Faculdade de Ciências, Universidade de Lisboa, P-1749-016 Lisboa, Portugal

Centro de Física Teórica e Computacional, Universidade de Lisboa, Avenida Professor Gama Pinto 2, P-1649-003 Lisboa, Portugal

H.J. Herrmann • J.S. Andrade Jr.
Computational Physics, IfB, ETH Zürich, Wolfgang-Pauli-Strasse 27, 8093 Zürich, Switzerland

Departamento de Física, Universidade Federal do Ceará, 60451-970 Fortaleza, Ceará, Brazil

© Springer International Publishing Switzerland 2016
A. Garas (ed.), *Interconnected Networks*, Understanding Complex Systems,
DOI 10.1007/978-3-319-23947-7_8

in unison. However, in our model orchestra, when listening to the musicians in the other sub-orchestra, the musicians in the same hall can split into two groups, playing with different tempi, leading to breathing synchronization. Interestingly, the partner in the other hall will be always playing with the same tempo but either in phase or anti-phase, depending on their tempo.

Understanding the consequences of a communication lag is also of major concern in other fields [7, 13, 16]. The plasmodium *Physarum polycephalum*, an amoeba-like organism consisting of a network of tubular structures for protoplasm flow, naturally shows periodic variations in its thickness, a useful skill when fighting predators. In a controlled experiment, two regions of the same organism have been physically separated by a certain distance with the possibility of fine tuning the communication between them [32, 33]. Depending on the coupling strength and time delay, the regions have been shown to present phase and anti-phase synchronization of the oscillatory thickness. This is precisely what we show here for the orchestra in the regime of strong influence of other musicians in the same concert hall. In what follows, we focus on the example of the orchestra but the results might also have impact on several biological and techno-social systems as, for example, functional brain networks, living oscillators, or coupled power grids. For a more general discussion of these findings see Ref. [21].

When playing the same piece, musicians in an orchestra try to synchronize their tempi, i.e., they try to play the notes at the same pace. Thus, for example, one violinist focuses simultaneously on the tempi of instruments in the same hall and on the corresponding violin in the other hall. All the other musicians act in the same way. The synchronizability of this setup can then be discussed in the framework of interconnected networks of oscillators. Each concert hall is modeled by a network, where nodes are musicians and links represent the interaction between them. Additionally, each musician also establishes a special inter-network coupling with one partner playing the same instrument in the other network (concert hall). Intra- and inter-network couplings have different time scales: while the intra-network interactions can be considered instantaneous, the inter-network ones have a time delay that depends on the distance between concert halls. Recent geometrical studies of coupled networks with intra- and inter-network links have revealed novel features never observed for isolated networks [14]. In particular, it has been shown that the overall robustness is reduced [28] and the collapse of the system occurs through large cascades of failures [8, 9]. Dynamic properties of coupled networks have also been studied [2, 10, 15, 19, 23, 30, 35].

The Kuramoto model is a usual approach to network synchronization [1, 3–6, 17, 18, 20, 22, 24–27, 31, 34]. For illustration purposes, here we stay with the example of an orchestra. A population Θ of n Kuramoto oscillators is considered to be mutually interacting, such as the musicians in one concert hall trying to keep the same tempo. We consider a random graph of average degree four. Each musician (oscillator) $i \in \Theta$ is described by a phase $\theta_i(t)$, representing her/his current position, and a natural tempo ω_i, corresponding to the pre-defined tempo of the music. Since all musicians are playing the same piece, we assume $\omega_i \equiv \omega_0$. The actual tempo of a musician is defined as the time derivative of the phase, $\dot{\theta}_i(t)$. To play the piece harmoniously, musicians try to synchronize their tempi. This interaction can be modeled in terms of the Kuramoto model as $\dot{\theta}_i = \omega_0 + \sigma \sum_{j=1}^{n} A_{ij}^{\Theta} \sin(\theta_j - \theta_i)$, where the sum goes over all other musicians ($i \neq j$), σ is the coupling strength between them, and $\mathbf{A}^{\Theta}$ is the connectivity matrix such that $A_{ij}^{\Theta} = 1$ if musician i is influenced by j and zero otherwise. For simplicity, we assume that the musicians are all playing at the same unitary amplitude, so that the state of each musician can be described by a phasor $e^{i\theta_i(t)}$.

The collective performance in one concert hall, namely, the synchronization of its network, is characterized here by the complex order parameter $r_{\Theta}(t)e^{i\Psi(t)} = \frac{1}{n} \sum_{j}^{n} e^{i\theta_j(t)}$, where the sum goes over all musicians, $\Psi(t)$ is the average phase, and the amplitude $0 \leq |r_{\Theta}(t)| \leq 1$ measures the global coherence, i.e., how synchronized the musicians are. If $r_{\Theta}(t) = 1$ all musicians play the same note at the same time, while low values of r_{Θ} imply that a significant fraction of musicians are out of phase.

We introduce now a second population Γ, also of n oscillators, representing the second concert hall. Within this network, musicians are coupled in the same way. Each $j \in \Gamma$ is coupled with the corresponding partner $i \in \Theta$, forming the inter-network couplings. In analogy to oscillators in Θ, the motion of each oscillator is described by a phasor $e^{i\gamma_j(t)}$, of phase $\gamma_j(t)$. The inter-network coupling is subjected to a time delay τ, corresponding to the time required for information to travel between concert halls [29]. Previous studies introduced time delay among oscillators of the same population [11, 36]. Here we consider the competition between an *instantaneous* intra-network and a *delayed* inter-network coupling. In a nutshell, the performance of each musician is described by,

$$
\begin{cases}
\dot{\theta}_i = \omega_0 + \sigma_{\text{EX}} \sin\left(\gamma_{j(i)}^{t-\tau} - \theta_i\right) + \sigma_{\text{IN}} \sum_{k=1}^{N} A_{ik}^{\Theta} \sin(\theta_k - \theta_i) \\
\dot{\gamma}_j = \omega_0 + \sigma_{\text{EX}} \sin\left(\theta_{i(j)}^{t-\tau} - \gamma_j\right) + \sigma_{\text{IN}} \sum_{k=1}^{N} A_{ik}^{\Gamma} \sin(\gamma_k - \gamma_j)
\end{cases}
, \tag{8.1}
$$

where the superscript $t - \tau$ indicates the instant when the phases are calculated, and σ_{EX} and σ_{IN} are the inter and intra-network couplings, respectively. In music, frequency is typically related with the note. However, here we assume that the properties of the note are not relevant. Instead, the usual natural frequency of the Kuramoto model (ω) corresponds to the tempo and, therefore, frequency and tempo will be used as synonymous.

For two interconnected networks of oscillators with time delay, a weak intra-network coupling, and random initial distribution of phases, two frequency communities emerge within the same network, each synchronized with its mirror in a breathing mode, as shown in Fig. 8.1b in which the position of each empty circle represent its phase. A frequency lock occurs within communities, which move in a cohesive fashion on the complex plane. Interestingly, pairs of nodes that are part of different networks oscillate with the same frequency but might be either

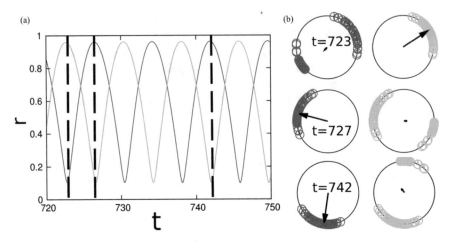

Fig. 8.1 The interactions between a strongly delayed inter-network coupling and a weak intra-network coupling create two communities of different frequencies in steady state. (**a**), The order-parameter of two populations (*red* and *green*) composed of $n = 100$ oscillators each with $\omega_0 = 1.0$, $\tau = 1.53$, $\sigma_{IN} = 0.01$, and $\sigma_{EX} = 0.5$. (**b**), Snapshot of populations at three different time steps (*black dashed vertical lines* in (**a**)), for the same parameters. Oscillators are spread across the complex plane according to their phase, with one plane for each network (*left* and *right columns*). The *arrow* at the center of each complex plane represents the order parameter of that network: its length representing the modulus and its argument representing the average phase. Superposition of the two groups of oscillators leads to breathing synchronization

in phase or anti-phase (phase shift of π). Consequently, the presence of these two frequency communities affects the measurement of the new global oscillatory state, called breathing synchronization [21]. Figure 8.1a shows the time evolution of the order parameters r_Θ and r_Γ for each population, quantifying this breathing behavior. For each curve, the maximum corresponds to the instant at which both groups of frequencies are in phase, while the minimum to an anti-phase between groups in the same network. Additionally, since for one frequency there is a phase shift of π between inter-network pairs of nodes, the minimum in one network corresponds, necessarily, to the maximum in the other. In the context of the orchestra this implies that the communication lag in the interaction between concert halls generates a tendency to split musicians in each sub-orchestra into two groups playing at different tempi, hindering harmony. Moreover, cohesion within each community affects the amplitude of the breathing resulting that the weaker the intra-network coupling, the smaller is this amplitude and more cacophony is present.

The observed breathing behavior is in deep contrast with what is expected for an isolated network ($\sigma_{EX} = 0$). For isolated networks, the classical Kuramoto model is recovered, with frequency and phase synchronization emerging at a critical coupling $\sigma_{IN} = \sigma_{IN}^*$. Above this threshold, a macroscopic fraction of oscillators synchronizes with the same frequency and since here we consider the same natural frequency for all oscillators ($\omega_i \equiv \omega_0$), $\sigma_{IN}^* \rightarrow 0$. The group of synchronized oscillators has frequency $\omega = \omega_0$ and the order parameter $r_\Theta(t)$ (or $r_\Gamma(t)$) saturates in time at a non-zero steady-state value, which is a monotonically increasing function of $(\sigma_{\text{IN}} - \sigma_{\text{IN}}^*)$ [17]. It is worth noticing that in the case of coupled networks, and for sufficient inter-network couplings, none of the two frequencies is ω_0.

Breathing synchronization is a direct consequence of the analytical solution obtained by Schuster and Wagner [29]. Depending on the initial phase difference between oscillators, the pair can synchronize with different frequencies ω, which are solutions of,

$$\omega = \omega_0 - \sigma_{\text{EX}} \sin(\omega\tau). \qquad (8.2)$$

Notwithstanding of oscillating with the same frequency in the stationary state, the two oscillators might either be in phase, if $\cos(\omega\tau) > 0$, or anti-phase. On inter-connected networks, in the limit $\sigma_{\text{IN}} = 0$, the stationary state incorporates all conceivable solutions of Eq. (8.2). Surprisingly, our results with a weak coupling uncover two frequency groups with phase locking. In any case, the observed frequencies are consistent with the solution of Eq. (8.2).

To summarize the impact of several combinations of parameters, we plot in Fig. 8.2 the phase diagram in the space of the two coupling strengths (σ_{IN} and σ_{EX}). To recognize each regime, we compute the amount of oscillators with consistent steady frequency below and above the mean value of possible frequencies over different samples. The color map of the main plot of Fig. 8.2 shows the ratio of these quantities. While the blue area represents the domain of σ_{IN} and σ_{EX} blends that leads to the smaller frequency, the shades in red represent the two regions where two frequencies can be accomplished. Notice however that the two synchronization regimes in red differ in their underlying features. The one in the left (lower σ_{IN}) is portrayed by the breathing behavior due to the presence of two frequency groups within each network. By contrast, in the supernode regime all nodes within a

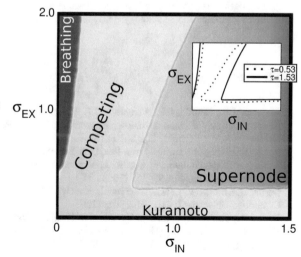

Fig. 8.2 Parameter space of two coupling strengths σ_{ex} and σ_{in} showing that the prevalence of one frequency over the other changes according to the coupling strengths. The color of each point represents the logarithm of the relative areas of the stable frequencies histogram. The *inset* exhibits the phase boundaries for different time delays. The dominant mechanisms of each region are labeled accordingly: breathing, (classical) kuramoto, competing, and supernode states (After [21])

network are in phase locking, with the same frequency and, in this way, the order parameter is steady in time at this state. In the inset, we show the phase boundaries for distinctive time delays from which one can similarly see that the transition between regimes changes considerably.

The presence of a time delay between two coupled networks of oscillators represents one more test to the worldwide control of an orchestra, or any other global system. We have shown that the interchange between coupling and delay leads to states of either a unique or two conceivable synchronized frequencies. We have discovered that, even with a weak intra-network coupling, oscillators inside the same network part into two frequency groups, each group mirroring another in the opposite network by oscillating at the same frequency. However, contingent upon their frequency, a group can be either in phase or anti-phase with its mirror in the other network, resulting in breathing synchronization. Additionally, we show that an arbitrary increment of the intra-network coupling is not an option to achieve phase and frequency synchronization regardless of its starting conditions. In a certain region of the parameter space, the intra-network coupling promotes the formation of two supernodes (one for every network), and two frequencies become stable.

It is safe to say that we are some years away of having a working setup of an intercontinental orchestra. In addition to all the technical challenges, our study shows that the very nature of a time-delayed system imposes a threat to the synchronization of the musicians. It would be rather interesting in fact to observe oscillations with two different frequencies within the musicians at the same continent due to the communication lag with the other region, resulting in a breathing synchronization cacophony.

Acknowledgements Authors would like to thank the Swiss National Science Foundation under contract 200021 126853, the CNPq, Conselho Nacional de Desenvolvimento Científico e Tecnológico - Brasil, the CNPq/FUNCAP Pronex grant, the ETH Zürich Risk Center, and the INCT-SC-Brasil for financial support. This work was also supported by grant number FP7-319968 of the European Research Council. NA acknowledges financial support from the Portuguese Foundation for Science and Technology (FCT) under Contract no. IF/00255/2013.

References

1. Acebrón, J.A., Bonilla, L.L., Vicente, C.J.P., Ritort, F., Spigler, R.: The Kuramoto model: a simple paradigm for synchronization phenomena. Rev. Mod. Phys. **77**, 137–185 (2005)
2. Araújo, N.A.M., Seybold, H., Baram, R.M., Herrmann, H.J., Andrade Jr, J.S.: Optimal synchronizability of bearings. Phys. Rev. Lett. **110**, 064,106 (2013)
3. Arenas, A., Díaz-Guilera, A., Kurths, J., Moreno, Y., Zhou, C.: Synchronization in complex networks. Phys. Rep. **469**, 93–153 (2008)
4. Barrat, A., Barthelemy, M., Vespignani, A.: Dynamical Processes on Complex Networks. Cambridge University Press, Cambridge (2008)
5. Boccaletti, S.: The Synchronized Dynamics of Complex Systems. Elsevier, Amsterdam (2008)
6. Boccaletti, S., Kurths, J., Osipov, G., Valladares, D.L., Zhou, C.S.: The synchronization of chaotic systems. Phys. Rep. **366**, 1–101 (2002)

7. Boccaletti, S., Latora, V., Moreno, Y., Chavez, M., Hwang, D.: Complex networks: structure and dynamics. Phys. Rep. **424**, 175–308 (2006)
8. Brummitt, C.D., D'Souza, R.M., Leicht, E.A.: Suppressing cascades of load in interdependent networks. Proc. Natl. Acad. Sci. U. S. A. **109**, E680–E689 (2011)
9. Buldyrev, S.V., Parshani, R., Paul, G., Stanley, H.E., Havlin, S.: Catastrophic cascade of failures in interdependent networks. Nature **464**, 1025–1028 (2010)
10. Cardillo, A., Gómez-Gardeñes, J., Zanin, M., Romance, M., Papo, D., Pozo, F.D., Boccaletti, S.: Emergence of network features from multiplexity. Sci. Rep. **3**, 1344 (2013)
11. Choi, M.Y., Kim, H.J., Kim, D., Hong, H.: Synchronization in a system of globally coupled oscillators with time delay. Phys. Rev. E **61**, 371–381 (2000)
12. Driessen, P.F., Darcie, T.E., Pillay, B.: The effects of network delay on tempo in musical performance. Comput. Music J. **35**, 76–89 (2011)
13. Duke, C.: Prosperity, complexity and science. Nat. Phys. **2**, 426–428 (2006)
14. Gao, J., Buldyrev, S.V., Stanley, H.E., Havlin, S.: Networks formed from interdependent networks. Nat. Phys. **8**, 40–48 (2011)
15. Gómez, S., Díaz-Guilera, A., Gómez-Gardeñes, J., Pérez-Vicente, C.J., Moreno, Y., Arenas, A.: Diffusion dynamics on multiplex networks. Phys. Rev. Lett. **110**, 028,701 (2013)
16. Helbing, D.: Systemic Risks in Society and Economics. Paper prepared for IRGC Workshop on Emerging Risks, Geneva (2009)
17. Kuramoto, Y., Nishikawa, I.: Statistical macrodynamics of large dynamical systems. Case of a phase transition in oscillator communities. J. Stat. Phys. **49**, 569–605 (1987)
18. Li, C., Chen, G.: Synchronization in general complex dynamical networks with coupling delays. Physica A **343**, 263–278 (2004)
19. Li, C., Sun, W., Kurths, J.: Synchronization between two coupled complex networks. Phys. Rev. E **76**, 046,204 (2007)
20. Louzada, V.H.P., Araújo, N.A.M., Andrade Jr, J.S., Herrmann, H.J.: How to suppress undesired synchronization. Sci. Rep. **2**, 658 (2012). doi:10.1038/srep00658
21. Louzada, V.H.P., Araújo, N.A.M., Andrade Jr, J.S., Herrmann, H.J.: Breathing synchronization in interconnected networks. Sci. Rep. **3**, 3289 (2013)
22. Lü, J., Chen, G.: A time-varying complex dynamical network model and its controlled synchronization criteria. IEEE Trans. Autom. Control **50**, 841–846 (2005)
23. Mao, X.: Stability switches, bifurcation, and multi-stability of coupled networks with time delays. Appl. Math. Comput. **218**, 6263–6274 (2012)
24. Motter, A.E., Zhou, C.S., Kurths, J.: Enhancing complex-network synchronization. Europhys. Lett. **69**, 334–340 (2005)
25. Néda, Z., Ravasz, E., Vicsek, T., Brechet, Y., Barabási, A.: Physics of the rhythmic applause. Phys. Rev. E **61**, 6987–6992 (2000)
26. Osipov, G., Kurths, J., Zhou, C.: Synchronization in Oscillatory Networks. Springer, New York (2007)
27. Pikovsky, A., Rosenblum, M., Kurths, J.: Synchronization: A Universal Concept in Nonlinear Sciences, vol. 12. Cambridge University Press, Cambridge (2003)
28. Schneider, C.M., Moreira, A.A., Andrade Jr, J.S., Havlin, S., Herrmann, H.J.: Mitigation of malicious attacks on networks. Proc. Natl. Acad. Sci. U. S. A. **108**, 3838–3841 (2011)
29. Schuster, H.G., Wagner, P.: Mutual entrainment of two limit cycle oscillators with time delayed coupling. Prog. Theor. Phys. **81**, 939–945 (1989)
30. Shang, Y., Chen, M., Kurths, J.: Generalized synchronization of complex networks. Phys. Rev. E **80**, 027,201 (2009)
31. Strogatz, S.H.: The Emerging Science of Spontaneous Order. Hyperion, New York (2003)
32. Takamatsu, A., Fujii, T., Endo, I.: Time delay effect in a living coupled oscillator system with the plasmodium of Physarum polycephalum. Phys. Rev. Lett. **85**, 2026 (2000)

33. Takamatsu, A., Takaba, E., Takizawa, G.: Environment-dependent morphology in plasmodium of true slime mold Physarum polycephalum and a network growth model. J. Theor. Biol. **256**, 29–44 (2009)
34. Wang, X.F.: Complex networks: topology, dynamics and synchronization. Int. J. Bifucart. Chaos **12**, 885–916 (2002)
35. Wu, X., Zheng, W.X., Zhou, J.: Generalized outer synchronization between complex dynamical networks. Chaos **19**, 013,109 (2009)
36. Yeung, M.K., Strogatz, S.H.: Time delay in the Kuramoto model of coupled oscillators. Phys. Rev. Lett. **82**, 648 (1999)

Chapter 9
Several Multiplexes in the Same City: The Role of Socioeconomic Differences in Urban Mobility

Laura Lotero, Alessio Cardillo, Rafael Hurtado, and Jesús Gómez-Gardeñes

Abstract In this work we analyze the architecture of real urban mobility networks from the multiplex perspective. In particular, based on empirical data about the mobility patterns in the cities of Bogotá and Medellín, each city is represented by six multiplex networks, each one representing the origin-destination trips performed by a subset of the population corresponding to a particular socioeconomic status. The nodes of each multiplex are the different urban locations whereas links represent the existence of a trip from one node (origin) to another (destination). On the other hand, the different layers of each multiplex correspond to the different existing transportation modes. By exploiting the characterization of multiplex transportation networks combining different transportation modes, we aim at characterizing the mobility patterns of each subset of the population. Our results show that the socioeconomic characteristics of the population have an extraordinary impact in the layer organization of these multiplex systems.

L. Lotero
Departamento de Ciencias de la Computación y de la Decisión, Universidad Nacional de Colombia, Medellín, Colombia
e-mail: llotero0@unal.edu.co

A. Cardillo
Laboratoire de Biophysique Statistique, École Polytechnique Fédérale de Lausanne (EPFL), CH-1015 Lausanne, Switzerland

Instituto de Biocomputación y Física de Sistemas Complejos, Universidad de Zaragoza, E-50018 Zaragoza, Spain
e-mail: alessio.cardillo@epfl.ch

R. Hurtado
Departamento de Física, Universidad Nacional de Colombia, Bogotá, Colombia
e-mail: rghurtadoh@unal.edu.co

J. Gómez-Gardeñes (✉)
Departamento de Física de la Materia Condensada, Universidad de Zaragoza, E-50009 Zaragoza, Spain

Instituto de Biocomputación y Física de Sistemas Complejos, Universidad de Zaragoza, E-50018 Zaragoza, Spain
e-mail: gardenes@gmail.com

© Springer International Publishing Switzerland 2016
A. Garas (ed.), *Interconnected Networks*, Understanding Complex Systems,
DOI 10.1007/978-3-319-23947-7_9

9.1 Introduction

Understanding human mobility patterns have attracted, for decades, the attention of researchers from many different scientific realms. The first models based on empirical observations date back to the 1940s and were elaborated by the sociologist Samuel A. Stauffer [1] and the philologist George K. Zipf [2]. It is at the end of the last century when, with the consolidation of mathematical frameworks such as the well-known *Gravity model* [3], when transportation science, as a realm of operations research, became a discipline on its own.

The advent of the big data era, have spurred the activity on transportation science and provided detailed datasets of real transportation systems. This characterization spans across many scales, from the short-range mobility patterns in urban areas [4–6] to world wide trips [7]. Remarkably, different degrees of resolution and types of information are nowadays available from the combined use of techniques for data gathering [8]. From the traditional datasets based on direct surveys [9], allowing to know the purpose of the trip (work/school, leisure, etc), to those large-scale ones gathered by tracking mobile communication systems [10, 11] or transport electronic cards [12]. This burst of activity have attracted many scientists from theoretical disciplines to contribute to the subject through the formulation of mobility models and mathematical tools aimed at reproducing and characterizing the observed patterns of movement [13–16].

The rapid change in the patterns of human mobility in the last decades, specially in what concerns the decrease in their duration together with the increase of their length, makes its characterization of utmost importance for many disciplines beyond the traditional scope of transportation science. The most paradigmatic example is the relevance of human mobility in the spread of diseases. The inclusion of the mobility ingredient into epidemic models has allowed to design sophisticated theoretical frameworks aiming at forecasting the onset and duration of pandemics with high time and spatial resolution [17–22].

In the last fifteen years, networks science [23–25] has appeared as the best suited mathematical frameworks to accommodate and characterize the interaction backbone of the very many complex systems captured by big data techniques. In fact, complex networks had been proposed as the natural framework to study spatially embedded systems [26] and, in particular, mobility networks. In these networks the different origins and destinations are represented as nodes of a graph, whereas the movements between locations are encoded as links connecting them [27]. Recently, thanks to the availability of more detailed information, it has been possible to represent many different types of transportation modes used for the movements within the same area under multilayer networks [28–30], in which each network layer represents a single transportation mode. In this way, each node still represents a particular origin/destination location and it is present in each of the network layers. However, links are represented in a different layer of interaction depending on the kind of transportation mode used for connecting two locations. This particular multilayer network is usually termed as *multiplex*.

In the recent years, different human mobility systems have been addressed under the paradigm of multiplex networks, ranging from urban movements [31] to medium [32] and large scale trips [33, 34]. Following this approach, here we address the multiplex structure of urban mobility in two different cities: Bogotá and Medellín. The novelty of the results presented rely on an additional ingredient of the mobility patterns that, up to our knowledge, has been ignored up to date. This new ingredient is the socioeconomic status *(SES)* of the individuals, mainly related to their wealth. Being the composition of many cities in the world highly hierarchical and inhomogeneous in terms of the capital distribution, it is thus relevant to unveil the influence that the different SES have on the mobility patterns.

To this aim, and considering that another relevant ingredient included in the available data sets is the transportation modes used by the individuals, we analyze the mobility patterns in terms of a multiplex network. In particular, we will analyze six different multiplex networks, each one corresponding to a different SES. Our approach relies on the *adiabatic projection* technique, introduced in [34], that consist in monitoring how the structural properties of the aggregate network show up as a result of the merging of the layers composing the multiplex. Thanks to this approach, it has been possible to spotlight how *segregation* and *multimodality* are characteristic of some particular social classes, and to unveil the dominant role played by the middle-class in the utilization of the transportation system as a whole.

The structure of this chapter is the following. We will first introduce the datasets used in Sect. 9.2 and the adiabatic projection technique together with the topological estimators in which it is used in Sect. 9.3. Section 9.4 is devoted to present the results of applying the former technique to the datasets of the cities of Bogotá and Medellín. Finally, in Sect. 9.5 we draw some conclusions and future work perspectives.

9.2 Urban Mobility and Socioeconomic Status

The mobility data presented and analyzed here are taken from surveys carried out in two major cities of Colombia: Bogotá and Medellín. These surveys were originally designed to collect information about travelers and their trips, so to identify traffic patterns and apply the results to urban and transportation planning. In these surveys, each householder is asked about the trips performed the day before the interview, providing with the origin and destination zones, the departure and arrival times, the transportation mode used and the purpose of each trip. In addition, householders are characterized by their socioeconomic characteristics, such as the age, gender, occupation, and the socioeconomic characteristic of their housing, which it is defined as its SES. The survey for the city of Bogotá, having a population of about 7 million of inhabitants, has a sample size of 45,446 people interviewed, reporting 100,846 trips [35]. On the other hand, the survey for the metropolitan area of Medellín, with a population of about 3.5 million people, reports 127,849 trips from 56,513 personal interviews [36]. However, not all the people interviewed made a trip and thus the number of travelers in both cities is smaller (see Table 9.1).

Table 9.1 Supplementary information about the mobility interviews. From left to right: number of travellers, P, their average age, $\langle A \rangle$, percentage of female subjects, Fem, total number of trips recorded, T_{TOT}, average number of trips per person, $\langle T \rangle$, average number of steps per trip, $\langle n_s \rangle$, number of urban areas (nodes), N, number of connections between areas (links), E

	P	$\langle A \rangle$	$Fem(\%)$	T_{TOT}	$\langle T \rangle$	$\langle n_s \rangle$	N	E
Bogotá	37,483	33.23	54.4	100,846	2.69	–	912	24,588
Medellin	45,496	33.34	53.4	127,849	1.58	1.105273	413	18,442

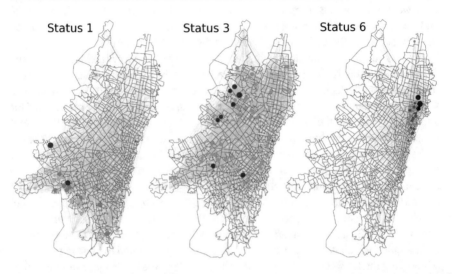

Fig. 9.1 Maps of the trips made in the city of Bogotá. Each map refers to a particular socioeconomic status, namely (from left to right) 1, 3 and 6. Each node corresponds to a different urban mobility zone, while a link (*light gray*) between two nodes indicates that a displacement has occurred among them. The size of the nodes indicates the amount of displacements (both from and to) occurring in that zone

In Table 9.1 we briefly show the most relevant information about the population interviewed in both cities. From the network perspective, the mobility graphs derived from these surveys contain $N = 912$ nodes (being both origins and destinations) for the city of Bogotá, and $N = 413$ for the metropolitan area of Medellín. In this way, two nodes are linked whenever the survey reports the existence of at least one trip between two zones. In addition, we take advantage of the socioeconomic information provided by the surveys, in particular the information about the SES of each individual, as this is a good proxy of the population wealth. This categorization of the population into strata is specific of Colombia, and ranges from status 1 for the lowest-income householders up to 6 for the highest-income individuals. Examples of mobility graphs of three of these socioeconomic groups are displayed in Fig. 9.1.

As introduced above, the aim of our work is to study the different means of transportations coexisting in the urban mobility as a multiplex mobility network. Since the surveys contained a number of different transportation means (25 in Bogotá and 17 in Medellín) we grouped these transportation modes into 6 different categories. In

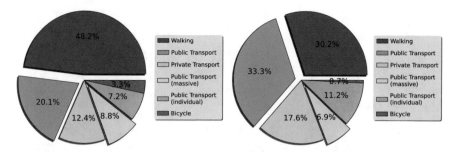

Fig. 9.2 Usage of the six transportation modes for the case of Bogotá (*left*) and Medellín (*right*)

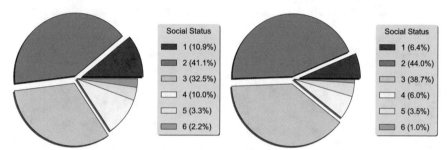

Fig. 9.3 Socioeconomic distribution of the population for the cities of Bogotá (*left*) and Medellín (*right*). SES ranging from 1 (poorest) to 6 (richest) is assigned to every traveler according to his/her economic wealth

particular: (i) pedestrian (walking), (ii) public transport, (iii) private transport (e.g. car or motorbike), (iv) public massive transport (e.g. metro), (v) public individual transport (e.g. taxi), and (vi) bicycle. The usage of each transportation group is displayed in Fig. 9.2. Surprisingly, being the same classification for both datasets, we notice that the usage of each group is not the same in both cities. This is due to several factors, such as the different morphology of the cities and the differences in their urban development and planning. Regarding the socioeconomic composition of the population we report in Fig. 9.3 the partition of the samples used in the surveys of Bogotá and Medellín in agreement with the real socioeconomic distribution of both cities, with the majority of the population being in SES 2 and 3.

Our goal in the next sections is to use the mobility and socioeconomic data provided by these surveys, to explain how the SES of individuals affect the mobility patterns. To illustrate how the different SES make use of the available transportation modes we show in Fig. 9.4 two different mobility matrices, **M**. The first type of matrix (top) shows how the usage of a transportation mean is partitioned into the different strata, whereas the second one shows how individuals of a particular SES use the different available modes. In both cases, the latter information is casted in 6×6 matrices whose entries $M_{t,s}$ correspond to, in the first case, the fraction of trips that individuals belonging to SES s perform using transportation mode t. In its turn, in the second case, the entry $M_{t,s}$ accounts for the probability that a trip of an

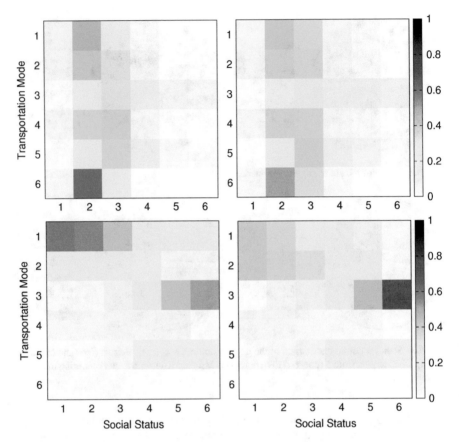

Fig. 9.4 Distribution of usage of transportation modes by the 6 SES. The matrices in the *left* (*right*) correspond to the city of Bogotá (Medellín). The matrices in the *top* (*bottom*) are normalized as $\sum_s M_{ts} = 1$ ($\sum_t M_{ts} = 1$). Transportation modes, from 1 to 6, are ordered according to Fig. 9.2

individual of SES s is performed using mean t. This information is shown for both cities, Bogotá (left) and Medellín (right).

At first sight, comparing those matrices in the left (Bogotá) and those in the right (Medellín), both cities display roughly the same usage patterns. In particular, concerning the matrices in the top we can observe that the usage of modes follows two different patterns depending on the precise transportation mean. For instance, modes 3 (private transport) and 5 (public individual transport) accumulate users from SES 2 to 6, whereas modes 4 (public massive transport) and 6 (bicycle) are mainly used by individuals with SES 2 and 3.

The second group of matrices (bottom) show, both for Bogotá and Medellín, that although multimodality is somehow present for individuals from SES 3 and 4, there is a high tendency to concentrate the trips around few transportation means. In fact, this concentration makes clear the socioeconomic differences according to the means selected: poor individuals from SES 1 and 2 concentrate their trips using modes 1 (pedestrian) and 2 (public transport), which are the cheapest ones, while those belonging to SES 5 and 6, mostly use means 3 (private transport) and 5 (public individual transport), that represent the most expensive ones. Thus, in the first case the concentration of trips around means 1 and 2 is due to the segregation of SES 1 and 2 towards cheap means whereas individuals from SES 5 and 6 can select their means according to their commodity. Thus, the mobility patterns in both cities show a clear transition segregation-multimodality-selection when going from the poorest to the richest.

9.3 The Adiabatic Projection of a Multiplex

From the analysis of the mobility matrices in Fig. 9.4, it becomes clear that the usage of the different transportation modes depends strongly on the status of the individuals. These results demand the analysis of how the different transportation modes are associated forming a *mobility multiplex network* (MMN) for each social status. In this section we present the *Adiabatic Projection* (AP) technique used to characterize MMN and the structural quantities under study.

Following the formalism introduced by Battiston et al. [41], we consider the MMN of a social status s as a system composed of N nodes and $M = 6$ layers. As explained before, nodes correspond to the different urban areas in a city. Layers, instead, represent different transportation modes. Keeping in mind such setup, and particularizing in the mobility multiplex of a given social status s, it is possible to associate to each layer α ($\alpha = 1, \ldots, M$) a graph $\mathcal{G}^{[s,\alpha]}(N, \mathcal{E}^{[s,\alpha]})$ described by an adjacency matrix $\mathbf{A}^{[s,\alpha]}$ whose entries are defined as $a_{ij}^{[s,\alpha]} = 1$ if zones i and j are connected by (at least) a trip of an individual from status s using transportation mode α. Under this formalism, the MMN of a social status s is fully described by the so-called *vector of adjacency matrices* $\mathbf{A}^s$ given by:

$$\mathbf{A}^s = \left\{ \mathbf{A}^{[s,1]}, \ldots, \mathbf{A}^{[s,M]} \right\} . \tag{9.1}$$

Once having introduced the basic notation characterizing each of the MMN, we describe the AP procedure used to study the coexistence of several interaction (here transportation) modes in a multiplex network. The technique relies in merging together a subset $V(m)$ containing $m \leq M$ layers into a single (monolayer) graph $\mathcal{G}^{[s,V(m)]}(N, \mathcal{E}^{[s,V(m)]})$ where:

$$\mathcal{E}^{[s,V(m)]} = \bigcup_{\alpha \in V(m)} \mathcal{E}^{[s,\alpha]} . \tag{9.2}$$

Therefore, the network $\mathcal{G}^{[s,V(m)]}$ is obtained by projecting all the layers contained in $V(m)$ onto a single one and by converting the multiple links (those existing in several layers in $V(m)$) into single ones. In this way, the topology of the resulting projected network is described by the *projected adjacency matrix* $\mathbf{A}^{[s,V(m)]}$ defined as:

$$a_{ij}^{[s,V(m)]} = \begin{cases} 1 & \text{if } \sum_{\alpha \in V(m)} a_{ij}^{[s,\alpha]} > 0 \\ 0 & \text{if } \sum_{\alpha \in V(m)} a_{ij}^{[s,\alpha]} = 0 . \end{cases} \tag{9.3}$$

The purpose of the AP of the layers of a multiplex is to analyze the evolution of some topological quantities when passing from single layers to the projected network resulting from merging all the M layers of the multiplex. Thus, the approach, introduced in [34] to study the *European Air Transportation Multiplex*, consists in varying the number of layers contained in the subset $V(m)$ from $m = 1$ to $m = M$. It is important to notice that the AP method (as introduced in [34]) considers, for each value m, the set $\mathcal{V}(m)$ containing all the possible subsets $V(m)$ comprising m layers. In this way, given a topological quantity x, one evaluates x in each projected graph $\mathcal{G}^{[s,V(m)]}$ derived from each subset $V(m)$ contained in $\mathcal{V}(m)$ and average the values obtained over all the resulting graph. Thus, given m, the average value of x in $\mathcal{V}(m)$ reads:

$$\langle x \rangle(m) = \frac{m!(M-m)!}{M!} \sum_{V(m) \in \mathcal{V}(m)} x(\mathcal{G}^{[s,V(m)]}) . \tag{9.4}$$

Note that although for $m = 1$ there are M possible subsets in $\mathcal{V}(1)$, for $m = M$ there is only 1 subset $V(M)$, while for a general value m the cardinality of $\mathcal{V}(M)$ can be extremely large. Thus, the AP technique demands a computationally expensive statistical treatment to cover all the possible layer combinations included in the sum of Eq. (9.4) when the number of layers M is large enough.

Here, instead, we get rid off the statistics over the sets $\mathcal{V}(m)$. In particular, based on the details contained in the dataset, we make use of the information about the usage of each transportation mode α by each SES s so that, for a certain value of m, we consider the projected graph $\mathcal{G}^{[s,V(m)]}$ constructed by merging the m most used transportation modes (layers) by SES s. In this way, for each value of m, $\mathcal{V}(m)$ contains one single subset $V(m)$ and thus we will denote each projected graph as $\mathcal{G}^{[s,m]}$ and its associated adjacency matrix as $\mathbf{A}^{[s,m]}$. Apart from the computational simplification of this variant of the AP technique, the new path from $m = 1$ to $m = M$ informs about how the individuals of a particular SES are benefited by adding transportation modes to their trips allowing to distinguish between strata displaying either segregation or selection of modes and those socioeconomic compartments showing multimodality.

The topological quantities studied with the AP technique cover traditional structural measures, used in simple networks, and others that take into account the layer structure of a multiplex.In particular, for each graph $\mathcal{G}^{[s,m]}$ we will study the following usual properties:

- *The size, S, of the giant component and the number of components, n_c.* It is important to note that S is normalized to be $0 \leq S \leq 1$, so that $S = 1$ when the N nodes in the network take part of a unique component. In addition, to compute n_c we have considered that isolated nodes do not constitute a component so that components contributing to n_c are those of size equal or larger than 2.
- *The average path length, L.* As usual, L is the average length of the shortest paths among all the couples of nodes in the network. Since the networks under study are highly disconnected, especially for small values of m, we have adopted the typical way out to avoid divergences in L, i.e., to consider only the nodes in the giant component.
- *The average degree, $\langle k \rangle$.* Again, in order to compute the average number of connections of the nodes we have excluded isolated nodes.
- *The clustering coefficient, C.* As usual, the clustering coefficient shows the probability that two nodes i and j having a common neighbor l are also connected. In this case also, isolated nodes do not contribute to clustering.

The above measures are those traditionally used for characterizing simple (single-layer) networks. However, there also exist measures that are specifically designed for multiplex networks (see the recent reviews [29, 30]). This is the case of the *Overlap, O.* The overlap quantifies the redundancy of links between layers, i.e., the fact that a link between two given nodes i and j is present in several layers. In our multiplex networks the existence of a large overlap would imply a large tendency of the individuals (belonging to the same SES) to use different means of transportation for connecting the same urban areas i and j. In the recent years, several overlap measures have been proposed [37–41]. Here, for a given value of the number of projected layers m, the overlap of the resulting graph $\mathcal{G}^{[s,m]}$ is measured via two different quantities, namely:

$$O_1 = \frac{W - K}{K}, \tag{9.5}$$

$$O_2 = \frac{D}{K}, \tag{9.6}$$

where K is the number of links in the aggregate graph $\mathcal{G}^{[s,m]}$, W is the total sum of the links in each of the m layers merged in $\mathcal{G}^{[s,m]}$, and D is the number of redundant links in the set of m layers. We can express these quantities, W, K and D, making use of the adjacency matrices associated to each layer, $\mathcal{G}^{[s,\alpha]}$, and that of the projection of the m most used layers, $\mathcal{G}^{[s,m]}$, as:

$$W = \sum_{\alpha=1}^{m} \sum_{i,j=1}^{N} a_{ij}^{s,\alpha} , \tag{9.7}$$

$$K = \sum_{i,j=1}^{N} a_{ij}^{s,m} , \tag{9.8}$$

$$D = \Theta \left(\sum_{\alpha=1}^{m} a_{i,j}^{s,\alpha} - 2 \right) , \tag{9.9}$$

where, in the last equation, $\Theta(z)$ is the step function defined as $\Theta(z) = 0$ for $z < 0$ and $\Theta(z) = 1$ otherwise.

9.4 Results

In this section, and relying on the AP technique of the MMN, our aim is to unveil the mobility patterns associated to the use of the transportation modes of each SES and, moreover, to monitor how the different patterns present in the transportation layers are combined into their corresponding mobility networks.

We start by studying how the combination of different transportation modes cover the different urban areas. This view can be explained as a percolation process driven by the addition of network layers (instead of nodes or links as in the traditional percolation contexts). To this aim, we focus on the evolution with m of the size of the giant connected component, $S(m)$, as well as the evolution for the number of components $n_c(m)$, and that of average path length of the giant component, $L(m)$. In Fig. 9.5, we show these evolutions for each of the 6 SES for the cases of Bogotá (top) and Medellín (bottom).

The adiabatic evolution of the giant component $S(m)$ shows that both cities behave in a similar way so that the different evolution $S(m)$ for the SES follows the same hierarchy. In particular, SES 2 and 3 reach to cover almost all the urban mobility zones of the cities. On the other hand, the coverage of SES 6 in both cities and also 5 in Bogotá are well below the 50 % of the zones. The main difference between the two cities shows up by looking at the rate $S(m)$ increases. While for Medellín the rate of change is very small for all the SES, in Bogotá, SES 1, 2 and 3 need to merge at least two different transportation layers in order to achieve the 80 % of their corresponding coverage. This result is the fingerprint of the segregation of these poor SES observed combined with the effect of the smaller sample size of the Bogotá survey (as compared to that of Medellín) that makes difficult to capture weak connections between urban mobility zones. These fact seems to affect more poor SES due again to their spatial segregation.

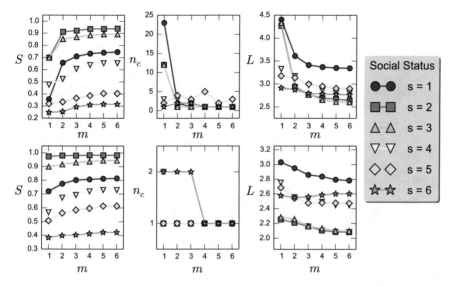

Fig. 9.5 Adiabatic evolution of the size of the giant component $S(m)$ (*left panels*), number of components $n_c(m)$ (*central panels*), and average path length L (*right panels*). The *top panels* are for the city of Bogotá, while those in the *bottom* refer to Medellín. Each curve corresponds to a different SES. Namely: 1 (●), 2 (■), 3 (▲), 4 (▼), 5 (◆) and 6 (★)

The evolution of the number of components $n_c(m)$, and the average path length $L(m)$ in the city of Bogotá further confirms the effects of the segregation of SES 1, 2 and 3. As observed, the initial ($m = 1$) values of both n_c and L are extremely large and they need to merge at least two transportation modes to reach small values of n_c and L. This is not the case for SES 4, 5 and 6 for which the evolution is far more smooth. Concerning the final values of L in the city of Bogotá, it is remarkable the large steady value reached by SES 1 as compared to the rest of the population. Thus, even if they can cover a large number of zones the trips connecting them associate in a rather linear way, thus not displaying shortcuts. In its turn, the situation in Medellín concerning the evolution of $n_c(m)$ is not pretty much like to that of Bogotá. In fact, in this city the locations of the usual destinations appear to be very clustered, leading to have a system composed of only one component even for $m = 1$ for most of the SES. The evolution of $L(m)$ instead is more interesting. As in the case of Bogotá, L decreases with m although in a smoother way [as occurred for the evolution of $S(m)$]. Again, it is worth to notice how, as in the case of Bogotá, SES 1 displays a different behavior from the rest of strata.

Summarizing, both the behavior of $S(m)$ and $L(m)$ point out that in both Bogotá and (more clearly) Medellín the six SES can be regrouped into three mobility compartments related with their wealth. Namely: low (SES 1), mid-low (SES 2 and 3) , and mid-high (SES 4, 5 and 6) compartments.

In Fig. 9.6, we confirm the above compartmentalization by monitoring the evolution for the number of different trips from/to each urban area (here represented

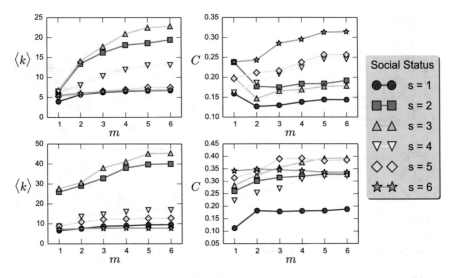

Fig. 9.6 Adiabatic evolution of the number of different destinations reached from a place, $\langle k \rangle$ and clustering coefficient, C as a function of the number of layers merged together, m. *Left panels* display the evolution for the average connectivity $\langle k \rangle (m)$, while those in the *right* show that of the clustering coefficient $C(m)$. The *top panels* are for the city of Bogotá, while those in the *bottom* refer to Medellín. Each curve corresponds to a different SES. Namely: 1 (●), 2 (■), 3 (▲), 4 (▼), 5 (♦) and 6 (★)

as the average degree $\langle k \rangle$ of the nodes) and the role of the various transportation modes in the triadic closure phenomenon, here studied via the clustering coefficient C. The evolution $\langle k \rangle$ vs. m reveals two clearly distinct behaviors. First, for SES 2 and 3 (also 4 in the city of Bogotá) incorporating transportation modes implies to increase the number of origins/destinations, pointing out the genuine multimodal character of these individuals who assign different transportation modes depending on the trip to be performed. On the other hand, for the rest of the SES there is almost no evolution. However, when looking back to the evolution of $S(m)$ in Fig. 9.5, it is easy to notice that the almost steady behavior of $\langle k \rangle (m)$ for these strata has different roots. While individuals belonging to SES 5 and 6 move from/to a limited amount of different places [as displayed by the small values of $S(m)$] using few transportation modes, due to the aforementioned selection mechanism, SES 1 displays a large coverage. Thus, for SES 1, the addition of a new transportation layer is mostly devoted to join pairs of disconnected nodes, and thus not used to increment the communication power of zones for which a trip already exists.

The particular way of evolution with m displayed by SES 1 is also related to the large resulting networks [as displayed by $L(m)$ in Fig. 9.5] and further confirmed by looking at the evolution of the clustering coefficient, $C(m)$. In both cities the values displayed by SES 1 are the smallest of the population and it does not show any significant change when increasing m. At variance, SES 5 and 6 display the largest values for the clustering in both cities, thus confirming again that, in these

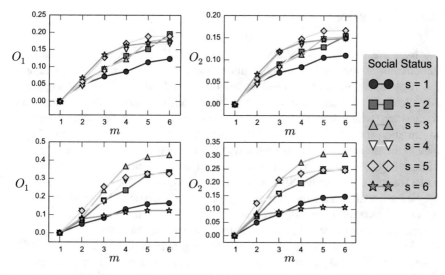

Fig. 9.7 Adiabatic evolution of the overlap as a function of the number of layers merged together, m. *Left panels* display the evolution for first definition of overlap, O_1, Eq. (9.5), while those in the *right* show the second one, O_2, Eq. (9.6). The *top panels* are for the city of Bogotá, while those in the *bottom* refer to Medellín. Each curve corresponds to a different SES. Namely: 1 (●), 2 (■), 3 (▲), 4 (▼), 5 (◆) and 6 (★)

cases individuals cover a limited and rather fixed number of zones, thus, favoring the formation of triadic paths in the aggregated graph.

Finally, in Fig. 9.7 we present the evolution for two measures of the overlap proposed above (see Eqs. 9.5 and 9.6). Interestingly, for the two cities the two measures present almost the same evolution in terms of the relative growth between the different SES. Considering the definitions of O_1 (that takes into account the total amount of degeneration of the links) and O_2 (that only count once the redundant links regardless of the number of times they are repeated) it is clear that $O_1 \geq O_2$. However, the similar trends observed and the small difference in the values attained by O_1 and O_2 point out that all of the SES do not tend to accumulate more than two overlapping links. Concerning the differences in the increase rates of O_1 and O_2 between SES we observe that, in both cities, individuals belonging to SES 1 tend to avoid overlapping, in agreement with the way (discussed before) SES 1 tend to increase the size of the giant component. Importantly, for the city of Bogotá the trends observed in both $O_1(m)$ and $O_2(m)$ seem to reproduce the three mobility compartments discussed above (1, $2 - 3$ and $4 - 5 - 6$) for both cities. However, the results for the city of Medellín are completely in disagreement with these compartments since, for instance, SES 6 display small overlapping tendency (being similar to that of SES 1) in contrast to the large tendency of SES 4 and 5.

9.5   Conclusions

We have presented a dataset about the human mobility in urban areas with two ingredients of utmost interest: the information about the multimodal nature of the trips, and the *socioeconomic status* of the individuals. The first ingredient has allowed us to tackle the analysis of the mobility patterns using the multiplex framework, which has attracted many attention lately. On the other hand, the information about SES provides with a novel ingredient that has been ignored up to date in the studies about human mobility. Exploiting these two ingredients, the aim of this chapter has been to describe how the different socioeconomic compartments make use and combine the different transportation layers.

We have analyzed the mobility multiplex networks of each SES by studying how different structural descriptors evolve as network layers (the transportation modes) are merged. This procedure, called adiabatic projection, starts from the network of the trips performed by means of the most used transportation mode and subsequently adds the layers corresponding to the other means in descending order of usage.

The main result of our work is the classification of the 6 SES into three compartments according to their behavior. Namely, in a first group we have SES 1 and 2, the poorest ones, whose behavior is characterized by the segregation, i.e., the usage of few and cheap transportation modes to cover a large fraction of the urban areas in a rather sparse way. The second compartment is composed of SES 3 and 4, having a genuine multimodal pattern and covering almost the total number of urban zones. Finally, the elite compartment composed of SES 5 and 6, is characterized by a selection of costly modes for performing the trips that, in their turn, display a very small coverage in terms of the urban areas reached although the connectivity within these zones turns to be rather dense.

The unveiled differences in the organization of the mobility multiplex networks according to SES demands the inclusion of this novel ingredient in the studies about human mobility and intrinsically related processes. As an example, it would be of interest to incorporate the presence of socioeconomic differences when studying the development of contagion processes in urban areas. We hope that our work will motivate more studies in this direction.

Acknowledgements We acknowledge financial support from the European Commission through FET IP projects MULTIPLEX (Grant No. 317532) and PLEXMATH (Grant No. 317614), from the Spanish MINECO under projects FIS2011-25167 and FIS2012-38266-C02-01, from the Departamento de Industria e Innovación del Gobierno de Aragón and Fondo Social Europeo (Grupo FENOL), and from the Universidad Nacional de Colombia under grants HERMES 19010 and HERMES 16007. JGG is supported by the Spanish MINECO through the Ramón y Cajal program. AC acknowledge the financial support of SNSF through the project CRSII2_147609. We thank *Area Metropolitana del Valle de Aburrá*, in Medellín, and *Secretaría Distrital de Movilidad*, in Bogotá, for the Origin-Destination Surveys Datasets.

References

1. Stouffer, S.A.: Intervening opportunities: a theory relating mobility and distance. Am. Sociol. Rev. **5**, 845–867 (1940)
2. Zipf, G.K.: The P_1P_2/D hypothesis: on the intercity movement of persons. Am. Sociol. Rev. **11**, 677–686 (1946)
3. Erlander, S., Stewart, N.: The Gravity Model in Transportation Analysis: Theory and Extensions. VSP, Utrecht (1990)
4. Batty, M.: The size, scale, and shape of cities. Science **319**, 769–771 (2008)
5. Porta, S., Latora, V., Wang, F., Rueda, S., Strano, E., Scellato, S., Cardillo, A., Belli, E., Cárdenas, F., Cormenzana, B., Latora, L.: Street centrality and location of economic activities in Barcelona. Urban Stud. **49**, 1471–1488 (2011)
6. Porta, S., Latora, V., Wang, F., Strano, E., Cardillo, A., Scellato, S., Iacoviello, V., Messora, R.: Street centrality and densities of retail and services in Bologna, Italy. Environ. Plan. B Plan. Design **36**, 450–465 (2009)
7. Guimerá, R., Mossa, S., Turtschi, A., Amaral, L.A.N.: The worldwide air transportation network: anomalous centrality, community structure, and cities' global roles. Proc. Nat. Acad. Sci. U. S. A. **102**, 7794–7799 (2005)
8. Asgari, F., Gauthier, V., Becker, M.: A survey on human mobility and its applications. arXiv:1307.0814 (2013)
9. Yan, X.-Y., Han, X.-P., Wang, B.-H., Zhou, T.: Diversity of individual mobility patterns and emergence of aggregated scaling laws. Sci. Rep. **3**, 2678 (2013)
10. González, M.C., Hidalgo, C.A., Barabási, A.-L.: Understanding individual human mobility patterns. Nature **453**, 779–782 (2008)
11. Wang, P., Hunter, T., Bayen, A.M., Schechtner, K., González, M.C.: Understanding road usage patterns in urban areas. Sci. Rep. **2**, 1001 (2012)
12. Roth, C., Kang, S.M., Batty, M., Barthélemy, M.: Structure of urban movements: polycentric activity and entangled hierarchical flows. PLoS ONE **6**, e15923 (2011)
13. Helbing, D., Buzna, L., Johansson, A., Werner, T.: Self-organized pedestrian crowd dynamics: experiments, simulations, and design solutions. Trans. Sci. **39**, 1–24 (2005)
14. Bazzani, A., Giorgini, B., Rambaldi, S., Turchetti, G.: Complexcity: modeling urban mobility. Adv. Complex Syst. (ACS) **10**, 255–270 (2007)
15. Song, C., Koren, T., Wang, P., Barabási, A.-L.: Modelling the scaling properties of human mobility. Nat. Phys. **6**, 818–823 (2010)
16. Simini, F., González, M.C., Maritan, A., Barabási, A.-L.: A universal model for mobility and migration patterns. Nature **484**, 96–100 (2012)
17. Eubank, S., Guclu, H., Kumar, V., Marathe, M.: Modelling disease outbreaks in realistic urban social networks. Nature **429**, 180–184, (2004)
18. Colizza, V., Barrat, A., Barthélemy, M., Vespignani, A.: The role of the airline transportation network in the prediction and predictability of global epidemics. Proc. Nat. Acad. Sci. U. S. A. **103**, 2015–2020 (2006)
19. Kleinberg, J.: Computing: the wireless epidemic. Nature **449**, 287 (2007)
20. Balcan, D., Colizza, V., Gonçalves, B., Hu, H., Ramasco, J.J., Vespignani, A.: Multiscale mobility networks and the spatial spreading of infectious diseases. Proc. Nat. Acad. Sci. U. S. A. **106**, 21484 (2009)
21. Tizzoni, M., Bajardi, P., Poletto, C., Ramasco, J.J., Balcan, D., Gonçalves, B., Perra, N., Colizza, V., Vespignani, A.: Real-time numerical forecast of global epidemic spreading: case study of 2009 A/H1N1pdm. BMC Med. **10**, 165 (2012)
22. Poletto, C., Tizzoni, M., Colizza, V.: Human mobility and time spent at destination: impact on spatial epidemic spreading. J. Theor. Bio. **338**, 41–58 (2013)
23. Albert, R., Barabási, A.L.: Statistical mechanics of complex networks. Rev. Mod. Phys. **74**, 47 (2002)

24. Newman, M.E.J.: The structure and function of complex networks. SIAM Rev. **45**, 167–256 (2003)
25. Boccaletti, S., Latora, V., Moreno, Y., Chavez, M., Hwang, D.: Complex networks: structure and dynamics. Phys. Rep. **424**, 175–308 (2006)
26. Barthélemy, M.: Spatial networks. Phys. Rep. **499**, 1–101 (2011)
27. Zaltz Austwick, M., O'Brien, O., Strano, E., Viana, M.: The structure of spatial networks and communities in bicycle sharing systems. PLoS ONE **8**, e74685 (2013)
28. De Domenico, M., Solé-Ribalta, A., Cozzo, E., Kivelä, M., Moreno, Y., Porter, M.A., Arenas, A.: Mathematical formulation of multilayer networks. Phys. Rev. X **3**, 041022 (2013)
29. Boccaletti, S., Bianconi, G., Criado, R., Del Genio, C.I., Gómez-Gardeñes, J., Romance, M., Sendiña-Nadal, I., Wang, Z., Zanin, M.: The structure and dynamics of multilayer networks. Phys. Rep. **544**(1), 1–122 (2014)
30. Kivelä, M., Arenas, A., Barthélemy, M., Gleeson, J.P., Moreno, Y., Porter, M.A.: Multilayer networks. J. Complex Netw. **2**(3), 203–271 (2014)
31. De Domenico, M., Solé-Ribalta, A., Gómez, S., Arenas, A.: Navigability of interconnected networks under random failures. Proc. Nat. Acad. Sci. U. S. A. **111**(23), 8351–8356 (2014)
32. Kurant, M., Thiran, P.: Layered complex networks. Phys. Rev. Lett. **96**, 138701 (2006)
33. Cardillo, A., Zanin, M., Gómez-Gardeñes, J., Romance, M., García del Amo, A.J., Boccaletti, S.: Modeling the multi-layer nature of the European air transport network: resilience and passengers re-scheduling under random failures. Eur. Phys. J. Spec. Top. **215**, 23–33 (2013)
34. Cardillo, A., Gómez-Gardeñes, J., Zanin, M., Romance, M., Papo, D., Del Pozo, F., Boccaletti, S.: Emergence of network features from multiplexity. Sci. Rep. **3**, 1344 (2013)
35. Secretaria Distrital de Movilidad: Informe de indicadores Encuesta de Movilidad de Bogotá 2011. Bogotá: Unión Temporal Steer Davies & Gleave Limited – Centro Nacional de Consultoría (2011). Retrieved from http://www.movilidadbogota.gov.co/?pag=1246
36. AREA Metropolitana del Valle de Aburrá: Capítulo 2: Diagnóstico. Formulación del Plan Maestro de Movilidad para la Región Metropolitana del Valle de Aburrá. Informe Final, pp. 21–72 (2006). Retrieved from: http://www.areadigital.gov.co/Movilidad/Documents/Plan%20Maestro%20de%20Movilidad.pdf
37. Barigozzi, M., Fagiolo, G., Garlaschelli, D.: Multinetwork of international trade: a commodity-specific analysis. Phys. Rev. E **81**, 046104 (2010)
38. Bianconi, G.: Statistical mechanics of multiplex networks: entropy and overlap. Phys. Rev. E **87**, 062806 (2013)
39. Kapferer, B.: Norms and the manipulation of relationships in a work context. In: Mitchell, J.C. (ed.) Social Networks in Urban Situations: Analyses of Personal Relationships in Central African Towns. Manchester University Press, Manchester (1969)
40. Parshani, R., Rozenblat, C., Ietri, D., Ducruet, C., Havlin, S.: Inter-similarity between coupled networks. Europhys. Lett. **92**, 68002 (2010)
41. Battiston, F., Nicosia, V., Latora, V.: Structural measures for multiplex networks. Phys. Rev. E **89**, 032804 (2014)

Chapter 10
The Weak Core and the Structure of Elites in Social Multiplex Networks

Bernat Corominas-Murtra and Stefan Thurner

Abstract Recent approaches on elite identification highlighted the important role of *intermediaries*, by means of a new definition of the core of a multiplex network, the *generalised K*-core. This newly introduced core subgraph crucially incorporates those individuals who, in spite of not being very connected, maintain the cohesiveness and plasticity of the core. Interestingly, it has been shown that the performance on elite identification of the generalised *K*-core is sensibly better that the standard *K*-core. Here we go further: Over a multiplex social system, we isolate the community structure of the generalised *K*-core and we identify the weakly connected regions acting as bridges between core communities, ensuring the cohesiveness and connectivity of the core region. This gluing region is the *Weak core* of the multiplex system. We test the suitability of our method on data from the society of 420,000 players of the Massive Multiplayer Online Game *Pardus*. Results show that the generalised *K*-core displays a clearly identifiable community structure and that the weak core gluing the core communities shows very low connectivity and clustering. Nonetheless, despite its low connectivity, the weak core forms a unique, cohesive structure. In addition, we find that members populating the weak core have the best scores on social performance, when compared to the other elements of the generalised *K*-core. The weak core provides a new angle on understanding the social structure of elites, highlighting those subgroups of individuals whose role is to glue different communities in the core.

B. Corominas-Murtra (✉)
Section for Science of Complex Systems, Medical University of Vienna, Spitalgasse 23, A-1090 Vienna, Austria
e-mail: bernat.corominas-murtra@meduniwien.ac.at

S. Thurner
Section for Science of Complex Systems, Medical University of Vienna, Spitalgasse 23, A-1090 Vienna, Austria

Santa Fe Institute, 1399 Hyde Park Road, Santa Fe, NM 87501, USA

IIASA, Schlossplatz 1, A-2361 Laxenburg, Austria
e-mail: stefan.thurner@meduniwien.ac.at

© Springer International Publishing Switzerland 2016
A. Garas (ed.), *Interconnected Networks*, Understanding Complex Systems,
DOI 10.1007/978-3-319-23947-7_10

10.1 Introduction

Which social network structures within a social system define an elite? Elites are typically formed from individuals that have the capacity to accumulate large amounts of wealth, power and influence. The location within the multiplex network of social interactions enables this small group of people to have significant influence and control over a large fraction of the population. A crucial feature of elites is that relations between its members define a highly cohesive network at different levels. Its defining traits are still under discussion [1–5]. Intuitively, elite structures are formed by individuals with a large number of ties connecting them to the overall society and by individuals who, in spite not being highly connected, link the highly connected ones. The later can be seen as *intermediaries* [6, 7].

A social system can be fairly described with a multiplex network (MPN) approach [8–10]. In a multiplex network, nodes interact through different types of relations or links. In this paradigm, elites have been thought to form a cohesive region which organises the whole topology of the multiplex system [11]. A few decades ago, quantitative sociology developed the concept of the K-core to identify this small subset of highly influencial individuals [12–14]. Generally members of the K-core tend to be highly connected (hubs). The strong-connectivity requirement in the definition of the K-core, docs not allow to identify the potentially important intermediaries or connectors. To improve this situation a *Generalised K-core* was suggested which includes connectors in the definition of the core of a complex network [7]. The suitability of this definition was demonstrated in a virtual society of players of the Massive Multiplayer Online Game (MMOG) *Pardus*, and was compared to the classic K-core for the identification of elites. The incorporation of connectors provides a much richer description of the core.

In this chapter we want to take the next logical step and analyse the substructure of elites. In particular we will focus on the weakly connected regions of the core, which provide the 'glue' for the different core communities. We expand the concept of a connector to an abstract structure which keeps the cohesiveness of the core of the multiplex network. The resulting subgraph we call the *weak core*, which defines the region of the core which prevents the core to disintegrate into its potential subcommunities. Interestingly, the notion of a weak core is independent of the definition of core and independent of the used community detection method.

We demonstrate our idea with data from the MMOG society of the game *Pardus* (http://www.pardus.at) [15], an open-ended online game with a worldwide player base which currently contains more than 420,000 people. MMOGs have been shown to be exceptional platforms over which quantitative results about social structures, dynamics, and organisational rules can be derived [7, 15–21]. In this game players live in a virtual, futuristic universe where they interact with other players in a multitude of ways to achieve their self-posed goals. A number of social networks can be extracted from the *Pardus* game, so that a dynamical multiplex network of a human social system can be quantitatively defined. The MPN consists of the time-varying communication, friendship, trading, enmity, attack, and revenge networks. Our findings in the virtual *Pardus* society confirm that indeed the weak core plays a crucial role in keeping the cohesiveness of the core of the multiplex system and show that members populating this subgraph are characterised by the largest scores in quantitative social performance indicators. The weak core might be a crucial and practical step towards the understanding of the internal structures of elites.

The chapter is organised as follows: In Sect. 10.2 we formally define the multiplex network, in Sect. 10.2 we revisit the concepts of *generalised K-core* and the *M-core*, which will be used as a community structure detector. Section 10.2.2 introduces the concept of the weak core. In Sect. 10.2.2 we discuss and define criteria to identify relevant levels of core organisation. Section 10.3 presents the results for the weak core analysis in the *Pardus* society. In Sect. 10.3 we discuss topological aspects, and in Sect. 10.3 the social performance indicators of those individuals in the weak core are compared to those comprising other social groups. Finally, in Sect. 10.4 we discuss the results.

10.2   Identification of the Weak Core

We introduce the following notation. We use bold letters for the various core subgraphs, namely **K**-core for the usual K-core subgraph, $\mathbf{G}_K$-core for the *generalised K-core*, **M**-core for the *M-core* and $\mathbf{MG}_K$ for the *M-core* of a generalised K-core. In general, we will use the word *core* to refer to the $\mathbf{G}_K$-core.

A multiplex system $\mathcal{M}$ is made of μ layers, which represent different types of interactions or relations among the same set of nodes. Nodes are usually people; for the multiplex we write

$$\mathcal{M} = \mathcal{M}(\mathcal{G}_1, \ldots, \mathcal{G}_\mu). \tag{10.1}$$

Levels or layers of the multiplex are indexes by greek letters. Figure 10.1 gives a schematic picture of the multiplex and the procedure described in the following.

Intersection of Levels in a Multiplex System

Each layer of the multiplex can be seen as a network $\mathcal{G}_\alpha(V, E_\alpha)$ whose set of nodes V is shared with the other layers $\mathcal{G}_1, \ldots, \mathcal{G}_\mu$ and whose set of links E_α describes the particular connections that occur at level α. The number of nodes of the multiplex system will be denoted by $|V|$ and the number of links of a given level α, $|E_\alpha|$. The *empty* graph, the graph with no nodes and no links, will be depicted by the symbol $\{\}$. The intersection graph $\mathcal{G}_\cap$ is defined as

$$\mathcal{G}_\cap = \bigcap_{\alpha \le \mu} \mathcal{G}_\alpha, \tag{10.2}$$

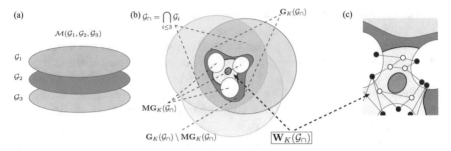

Fig. 10.1 A given multiplex system is composed by three layers, $\mathcal{G}_1$, $\mathcal{G}_2$ and $\mathcal{G}_3$ (**a**). Extracting the weak core (**b**): First, compute the intersection among the layers, $\mathcal{G}_\cap$, then compute the G_K-core of $\mathcal{G}_\cap$, namely $G_K(\mathcal{G}_\cap)$, depicted as the *red* region containing both *grey* and *white* components. After that we extract the **M**-core of the $\mathbf{G}_K$-core ($M = 2$) thereby obtaining the subgraph $\mathbf{MG}_K(\mathcal{G}_\cap)$, whose components are shown as white regions at the core. These three regions depict the communities defined through a high degree of clustering; we call them the *core communities*. The *weak core* of $\mathcal{G}$, $\mathbf{W}_K(\mathcal{G})$ (*grey region*), is the subgraph formed by all links and nodes that start in one of the core communities and end in another core community (*grey region*). No links between members of the same core community are allowed in $\mathbf{W}_K(\mathcal{G})$. In (**c**) we show some examples of potential structures forming the $\mathbf{W}_K(\mathcal{G})$-core. We differentiated the nodes belonging to $\mathbf{MG}_K(\mathcal{G}_\cap)$ (*black*) and to $\mathbf{G}_K(\mathcal{G}_\cap) \setminus \mathbf{MG}_K(\mathcal{G}_\cap)$ (*white*), to emphasise the hybrid, glue-like character of the weak core

where the intersection symbol means

$$\bigcap_{\alpha \leq \mu} \mathcal{G}_\alpha \equiv \mathcal{G}(V_\cap, \bigcap_{\alpha \leq \mu} E_\alpha). \tag{10.3}$$

Here $V_\cap$ is the set of nodes which are at the endpoint of at least to one link in $\bigcap_{\alpha \leq \mu} E_\alpha$. Nodes that become isolated after the intersection operation are not considered for any of the computations involving $\mathcal{G}_\cap$. Note that the more levels the multiplex has, the more probable is that $|V| > |V_\cap|$. One can of course intersect only specific layers of the multiplex. For the intersection of layers $\alpha_1, \ldots, \alpha_k$ we write for the intersection graph

$$\mathcal{G}_\cap^{\alpha_1, \ldots, \alpha_k} = \bigcap_{\alpha_1, \ldots, \alpha_k} \mathcal{G}_{\alpha_k}.$$

Links in a given intersection graph are referred to as *multi-links* [22]. In $\mathcal{G}_\cap^{\alpha_1, \ldots, \alpha_k}$, two nodes are linked if they are linked in layers $\alpha_1, \ldots, \alpha_k$. Links in $\mathcal{G}_\cap$ depict pairs of nodes which are connected through all the possible relations that define the multiplex – see Fig. 10.1a,b.

The G_K-Core and Its Community Structure

10.2.1 The G_K-Core

In the following we work with an intersection graph with layers that are considered relevant, for which we write $\mathcal{G}_\cap$. We then compute its *generalised K*-core, G_K-core, which is defined as the maximal induced subgraph for which each node has either a degree equal or larger than K, *or* it connects two nodes whose degree is equal or larger than K. Recall that, as for the **K**-core, the connectivity requirements must be satisfied inside the subgraph, so that a recursive algorithm must be used. The algorithm may work as follows: Starting with graph $\mathcal{G}$ we remove all nodes $v_i \in \mathcal{G}$ satisfying that: (1) its degree is lower than K *and* (2) at most one of its nearest neighbours has degree equal or higher than K. We iteratively apply this operation over $\mathcal{G}$ until no nodes can be pruned, either because the derived subgraph is empty, or because all nodes which survived the iterative pruning mechanism cannot be removed following the above instructions. The graph obtained after this process is the *generalised K*-core subgraph, referred to as G_K-core. The inclusion of the connectors in the definition of the G_K-core makes it a richer topological object. It has been shown that G_K is better suited for the identification of the *elite* in a social system than the standard $\mathbf{K} - core$ [7].

10.2.2 The M-Core and the Community Structure in the Core

The G_K-core can have internal structure itself around core communities. We assume that *core communities* are formed by regions of the core which are highly clustered. The identification of highly clustered regions is performed by means of the **M**-core [23]. Given a graph $\mathcal{G}$, the **M**-core of this graph, $M(\mathcal{G})$, is defined as the maximal induced subgraph of $\mathcal{G}$, in which *each link participates in at least M triangles*. The **M**-core highlights the role of triadic-closure within social dynamics, a process that seems to be a major driving force in social network formation [9, 24–27]. In our case we will use it to identify the clustered regions of $G_K(\mathcal{G}_\cap)$, which we denote by $MG_K(\mathcal{G}_\cap)$. Larger and lower values of M will identify more or less clustered regions in the core, respectively. The different connected components of $MG_K(\mathcal{G}_\cap)$ are the *core communities*.

Finally, we point out that the identified communities will in general not contain all the links associated with the core; also some nodes may be removed in the process. Formally this means that $G_K(\mathcal{G}_\cap) \setminus MG_K(\mathcal{G}_\cap) \neq \{\}$. This property will be relevant for the computation of the *Weak* core.

The Weak Core and the Minimal Weak Core

The *Weak core* is the subgraph of the core in which all nodes and links participate in a path that goes from one core community to another, without crossing any of such communities. The weak core, thus, ensures the cohesiveness of the core of the network, acting as a gluing structure between core communities.

We put the above informal statement in a more formal way, assuming the definitions of core and core community based upon the $\mathbf{G}_K$-core and $\mathbf{M}$-core, respectively. Let us assume that the core defined by $\mathbf{G}_K(\mathcal{G}_\cap)$ contains a single connected component and that the $\mathbf{MG}_K(\mathcal{G}_\cap)$ identifies several core communities $C_1, C_2, \ldots, C_m$ – which are, as we said above, the connected components of the $\mathbf{MG}_K$-core. The *weak core* of a multiplex graph, $\mathbf{W}_K(\mathcal{G}_\cap)$, is formed by all links and nodes of $\mathbf{G}_K(\mathcal{G}_\cap)$ that participate in a path that starts at some node $v_k \in C_i$ and ends at some $v_\ell \in C_j$, where C_i and C_j are different components of $\mathbf{MG}_K(\mathcal{G}_\cap)$, with the constraint that all nodes in the paths but v_k and v_ℓ, if any, must belong to $\mathbf{G}_K(\mathcal{G}_\cap) \backslash \mathbf{MG}_K(\mathcal{G}_\cap)$ – see Fig. 10.1b,c. The *Weak core* of a multiplex network is thus the region of the core of the intersection network which ensures the cohesiveness of the core. By definition the weak core itself is a weakly clustered region of the core, and its nodes may be among the least connected nodes of the core. In Fig. 10.1b,c we schematically show how such subgraphs can be derived.

We additionally define the *minimal weak core*, $\tilde{\mathbf{W}}_K$, as those links and nodes participating in all *minimal paths* from one component to an other in $\mathbf{MG}_K(\mathcal{G}_\cap)$. If there are two (or more) paths of $\mathbf{W}_K$ that connect $v_k \in C_i$ and $v_j \in C_x$, where $x \neq i$, we take the shortest. In case two or more paths that connect such two nodes have the minimal length, we choose one at random. Note that by construction, if $\mathbf{W}_K \neq \{\}$, then:

$$\mathbf{W}_K(\mathcal{G}_\cap) \cap \mathbf{MG}_K(\mathcal{G}_\cap) \neq \{\} \text{ and}$$

$$\tilde{\mathbf{W}}_K(\mathcal{G}_\cap) \cap \mathbf{MG}_K(\mathcal{G}_\cap) \neq \{\}.$$

The concept of the weak core is not tied to a particular definition of the *core* or a *core community*. One can define the core of a network in any suitable way (for example using the K-core). If it is possible to identify more than one community inside this core (using any method of community detection) the weak core is the region (links and nodes) that glues the communities. The reason by which we suggest the combination of the $\mathbf{G}_K$-core and the $\mathbf{M}_K$-core is that the first has been shown to perform better in identifying relevant levels of core organisation (especially in social systems) than the classical $\mathbf{K}$-core, and because the $\mathbf{M}$-core captures clustering. It may happen that the $\mathbf{W}_K$-core is composed of a set of links that connect different components of the $\mathbf{MG}_K(\mathcal{G}_\cap)$, thereby indicating that all nodes in $\mathbf{G}_K(\mathcal{G}_\cap)$ are in $\mathbf{MG}_K(\mathcal{G}_\cap)$, and that the $\mathbf{M}$-core extraction only removed a few links. Finally, we say the Weak core is empty if the application of the $\mathbf{M}$-core does not identify the communities within the $\mathbf{G}_K$-core.

Identifying Relevant Levels of Core Organisation

Which value of K should be used to compute the $\mathbf{G}_K$-core such that the *weak core* reveals significant topological information of the organisation of the multiplex? Informally speaking, if the $\mathbf{MG}_K$-core identifies a very large community and a set of other small communities, the role of the weak core will be less relevant than in the case where the communities, even eventually lower in number, have comparable size. The more uniform the size of the core communities, the more *relevant* are the level(s) for the core organisation.

To identify such level(s), we compute the $\mathbf{MG}_K(\mathcal{G}_\cap)$ for all values of K by which $\mathbf{MG}_K(\mathcal{G}_\cap) \neq \{\}$. For each of this levels, we proceed as follows: Let $C_1, C_2, \ldots, C_m$ be the m core communities of the $\mathbf{G}_K$-core, glued in this latter subgraph by means of the Weak core $\mathbf{W}_K(\mathcal{G}_\cap)$. Let $|C_i|$ be the number of nodes of the component M_i and let us define the probability that a randomly chosen node from $\mathbf{MG}_K(\mathcal{G}_\cap)$ is in the component C_j

$$p(C_j) = \frac{|C_j|}{\sum_{i \leq m} |C_i|}.$$

We then compute the corresponding Shannon entropy

$$H(\mathbf{MG}_K) = - \sum_{i \leq m} p(C_i) \log p(C_i). \tag{10.4}$$

The more uniform is the size distribution of the core communities, the larger will the entropy be. This enables us to compare different core community structures with the same number of components but with different community distribution sizes. For example, one can compare the situations where the $\mathbf{W}_K$-core glues two components of sizes 10 and 100, or 50 and 50. The role of the Weak core will be much more relevant within the core organisation in the second case than in the first one, and this is identified by the above entropy. To correct for size effects, we use the normalised Shannon entropy

$$h(\mathbf{MG}_K) = \frac{H(\mathbf{MG}_K)}{\log m}. \tag{10.5}$$

The most relevant level of core organisation, $K^\star$, if there exists any, will be located at the level K for which

$$h(\mathbf{MG}_K) = \max_K \{h(\mathbf{MG}_K)\}. \tag{10.6}$$

If such a level exists, this will define the optimal value of K with which the weak core will be computed.

Concerning the choice of M, in the computation of the $\mathbf{MG}_K$-core, we use the following observation: If a given core does not break at low values of M, this means that the core is highly clustered and highly cohesive. In terms of the core organisation, the role of the community structure (if any) will be less significant. We therefore choose M as the minimum value that breaks the $\mathbf{G}_K$-core. Generally we will consider $M > 1$, since values of $M = 1$ can only isolate regions with low clustering and can not capture the idea of cohesive community. One can use other levels of M to gain a better insight in the core structure of the graph.

10.3   Results

We demonstrate the feasibility and quality of identifying the 'connector regions' within the core of multiplex social systems with data from a social multiplex network of social interactions occurring in the virtual society of the *Pardus* computer game. The multiplex network is composed of cooperative interactions *friendship* (*F*), *communication* (*C*) and *trade* (*T*). Our social system is therefore given by the MPN $\mathcal{M}(t) = \mathcal{M}(V, E_F \times E_C \times E_T, t)$, where E_F, E_C and E_T are the sets of links defining a friendship relation, a communication event, or a commercial relation, respectively. Our analysis is performed on the three networks $\mathcal{G}_F, \mathcal{G}_C$ and $\mathcal{G}_T$ obtained from the most active players in two time windows of 60 days in length, $t_1 = 796\text{--}856$ and $t_2 = 1140\text{--}1200$. The time units here are days since beginning of the game. A link between two players in layer $\mathcal{G}_F$ exists if at least one player recognises the other as a 'friend' within a time window. A link between two players in layer $\mathcal{G}_C$ exists if at least one player has sent a message to the other, and a link between two players in $\mathcal{G}_T$ means that there has taken place at least one commercial transaction between the players in the time window. The set of players that defines the set V of the MPN obtained from the period 796–856 contains 2422 players, and 2059 players for the period 1140–1200. Inactive players are removed from the MPN which leaves us with about 2000–2500 players. Following Eq. (10.3) and with these players we construct,

$$\mathcal{G}_\cap = \mathcal{G}_F \bigcap \mathcal{G}_C \bigcap \mathcal{G}_T.$$

We drop the time label T indicating the time window. All results are presented for $\mathcal{G}_\cap$. Single layer analysis or even intersections of two layers show much more noisy and unclear trends. $\mathcal{G}_\cap$ allows us to use the multiplex structure to reduce noise.

Topological Indicators

In Fig. 10.2 we show the normalised entropy $h(\mathbf{MG}_K)$, Eq. (10.6), as a function of K for both time windows in (a) and (b), respectively. The value of $h(\mathbf{MG}_K)$ remains almost constant with a slight increase before it abruptly jumps to zero. This constant plateau – see Fig. 10.2a,b – is observed regardless if the number communities in the core of the network – see Fig. 10.2c,d. It is true even the number of communities has significant variations – see Fig. 10.2c,d. The number of communities shows a decreasing trend until only one community is identified, provoking the collapse of $h(\mathbf{MG}_K)$ to zero. Note that the collapse occurs just after the value of K at which the communities of the core have comparable size. If only a single layer

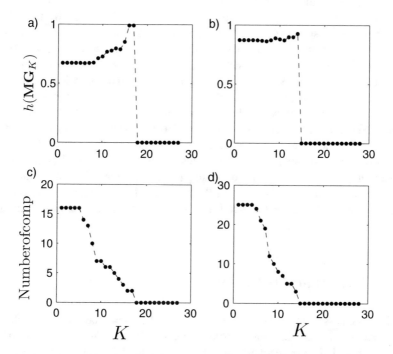

Fig. 10.2 Evolution of the relevance parameter $h(\mathbf{MG}_K)$ as a function of K for the two periods under study; 796–856 (**a**) and 1140–1200 (**b**). In both periods we observe a remarkably constant behaviour with a slightly increasing trend followed by an abrupt decay. The larger is $h(\mathbf{MG}_K)$, the more relevant are the weak structures keeping the core connected. In (**c**) and (**d**) we plot the raw number of core communities of the cores of the two periods under study against K. We observe that such number decreases over time, although the increase on $h(\mathbf{MG}_K)$ tells us that the breaking is more and more uniform as long as K increases, in terms of community sizes. Finally, the abrupt decay in $h(\mathbf{MG}_K)$ coincides with the fact that only one core community is identified, which occurs at deep levels of the core organisation. The K level in which the $\mathbf{W}_K$-core is computed is the one displaying the maximum $h(\mathbf{MG}_K)$. In the first period, (**a**), the $\mathbf{G}_K$-core is already broken at the levels showing the maximum normalised entropy ($K = 14, 15$), we thus choose the largest $h(\mathbf{MG}_K)$ by which the $\mathbf{G}_K$-core is not broken, $K = 13$

of the multiplex system is used, the situation is less well pronounced than the case shown in Fig. 10.2a,b. Relevant levels identified using the procedure described in Sect. 10.2.2 are found for $K = 14$ for the first period, 796–856 days, and $K = 13$ for the second, 1140–1200 days. Although for the first period $h(\mathbf{MG}_K)$ is higher for $K = 15$ and 16, at these stages the $\mathbf{G}_K$-core is already broken into two components, whereas at $K = 14$ it still contains one single component, as required by the proposed method.

To compute the $\mathbf{MG}_K$-core we set $M = 2$. The $\mathbf{MG}_K$-core detects three highly clustered communities of comparable size in both periods, containing 68 % and 61 % of all nodes of the $\mathbf{G}_K$-core in the first and second time period, respectively. These communities show a high clustering coefficient $c \sim 0.6$–0.7 (clustering of the $\mathbf{G}_K$-cores is $\sim$0.5), and an average degree of around $\langle k \rangle \sim 7$, which is similar to the average degree of the $\mathbf{G}_K$-cores in both periods. The relative sizes of the identified weak cores in relation to the respective $\mathbf{G}_K$-cores are 0.27 and 0.28 for the first and second periods, respectively. The $\mathbf{W}_K$-core is formed by a weakly connected region exhibiting less than $1/2$ of the average degree of the $\mathbf{G}_K$-core, $\langle k \rangle_{\mathbf{W}} = 3.0$ and $\langle k \rangle_{\mathbf{GK}} = 7.0$, and $\langle k \rangle_{\mathbf{W}} = 2.9$ and $\langle k \rangle_{\mathbf{GK}} = 6.8$ in two periods, respectively. As expected the clustering is almost vanishing around $c \sim 0.07$ in both periods.

The most surprising topological property of the observed weak cores is that, in spite their low connectivity and their role as connector regions, they define a single connected component in both time periods. This reveals that the $\mathbf{W}_K$-core plays an important functional role in the underlying organisation of the network. We find that in both time periods $\tilde{\mathbf{W}}_K \approx \mathbf{W}_K$. This means that the raw $\mathbf{W}_K$-core is quite optimal in the sense that a few redundant paths connecting the communities of the $\mathbf{G}_K$-core are identified. This confirms the property of the identified weak core as a true minimal gluing region that keeps the cohesiveness of the core of the multiplex network.

Social Performance Indicators

Social indicators and social performance measures of those players that populate the weak core show interesting and unexpected results. These indicators are: *Experience* is a numerical indicator accounting for the experience of the player, related to battles in which the player has participated, or the number monsters he/she killed. *Activity* is a numerical indicator related to the number and complexity of actions performed by the player. *Age* is the number of days after the player joined the game. Finally, *wealth* is a numerical indicator accounting for the wealth of the player within the game at any point in time. Wealth accounts for money and the cumulative value of a payers's equipment within the game. We list the average experience level, activity level, age and wealth of those nodes in Table 10.1.

The most salient observation is that for almost all indicators in the two periods under study, those nodes that compose the weak core have the highest social scores when compared to nodes composing the core, its clustered communities,

Table 10.1 Table with the social indicators of the different subgraphs of $\mathcal{G}_\cap$ corresponding to periods $t_1 = 796$–856 and $t_2 = 1140$–1200. Note that for the first period, $\tilde{\mathbf{W}}_K = \mathbf{W}_K$

Subgraph	⟨Experience⟩	⟨Activity⟩	⟨Wealth⟩	⟨Age⟩
t_1				
$\mathbf{G}_K$	4.9×10^5	3.63×10^6	5×10^7	677
$\mathbf{MG}_K$	4.77×10^5	3.62×10^6	4.88×10^7	668
$\mathbf{W}_K$	$\mathbf{6.01 \times 10^5}$	$\mathbf{4.11 \times 10^6}$	$\mathbf{5.18 \times 10^7}$	732
$\tilde{\mathbf{W}}_K$	$\mathbf{6.01 \times 10^5}$	$\mathbf{4.11 \times 10^6}$	$\mathbf{5.18 \times 10^7}$	732
t_2				
$\mathbf{G}_K$	7.72×10^5	5.69×10^6	9.84×10^7	1020
$\mathbf{MG}_K$	8.58×10^5	6.14×10^6	$\mathbf{1.13 \times 10^8}$	1060
$\mathbf{W}_K$	1.02×10^6	6.3×10^6	9.85×10^7	1030
$\tilde{\mathbf{W}}_K$	$\mathbf{1.03 \times 10^6}$	$\mathbf{6.38 \times 10^6}$	1.04×10^8	1070

or the average player. Even the communities of the core are defined by a strong connectivity pattern, which does not guarantee the best performance in social indicators. This tells us that being located between different core communities leads to superior social performance. We find one exception where the wealth in the second period is higher for the core communities. In addition, one finds that the age of the players populating the weak core is sensibly larger than the average age of the core and, in particular, larger than the average age of the core communities. In Table 10.1 we collect the results, highlighting the best scores. We finally note that in the second period the $\mathbf{MG}_K$-core is already broken into three components C_1, C_2, C_3 for $M = 1$. Remarkably, the weak core is formed only by two links, that connect C_1 with C_2, and C_1 with C_3. This identifies what one could call *supercritical links* at the core of the multiplex society.

10.4   Discussion

In this chapter we described a new type of subgraph, the weak core, which belongs to the family of core subgraphs. The latter include the clique subgraphs [28], the *Rich club* [29], the standard **K**-core [12–14], and the generalised **K**-core [7] as well as other approaches [30, 31]. The interest of this weak core arises since it captures a property that is essential for the identification of elite structure in social systems: The ability of the high social performers to maintain ties to the various core communities that organise the whole topology of the system from its core. The core of the multiplex network, defined as the generalised K-core of the intersection network from all layers in the MPN provides a rich structure in which one can identify core communities. In our case, we identified the community structure of the core of the MPN through the application of the **M**-core. In doing so, we consider that core communities are defined by those regions of the core which depict a highly clustered structure. In a totally opposite way, the weak core is comprised of regions of the core that are neither highly connected nor well clustered. This region's primary role is to keep the cohesiveness of the core.

The weak core identifies those individuals performing best in the virtual society. In previous studies, it has been shown that there is a direct relation between the degree of the player and its performance [21]. However, our findings indicate that nodes that are high social performers, well connected and part of a core group, need ties to other communities in the core. The weak core suggests a deeper structure of elites in social systems, and includes what seems to be a crucial for elite members: the ability to maintain ties beyond the community they belong to. Moreover, some members of the weak core may not belong to any core community and their role within the core organisation is purely devoted to keep the cohesive nature of it. This role as topological hinge between core communities may lead this particular class of players to an increase of their social performance.

The presented methodology is not tied to the particular definitions of the core or core community. Further works should stress the functional role of these weakly connected regions at the core of multiplex systems. In addition, the notion of weak core could be applied to other fields where this type of brokerage structure may play an important role in organising networks, such as in neurology or biological networks.

Acknowledgements This work was supported by the Austrian Science Fund FWF under KPP23378FW, the EU LASAGNE project, no. 318132 and the EU MULTIPLEX project, no. 318132.

References

1. Mills, C.W.: The Power Elite. Oxford University Press, Oxford (1956)
2. Mills, C.W.: The structure of power in american society. Br. J. Sociol. **9**(1), 29–41 (1958)
3. Keller, S.: Beyond the Ruling Class. Strategic Elites in Modern Society. Random House, New York (1963)
4. William, F.G.: Who Rules America? McGraw-Hill, New York (1967)
5. Bottomore, T.: Elites and Society, 2nd edn. Routledge, London (1993)
6. Friedkin, N.E.: Structural cohesion and equivalence explanations of social homogeneity. Sociol. Methods Res. **12**, 235–261 (1984)
7. Corominas-Murtra, B., Fuchs, B., Thurner, S.: Detection of the elite structure in a virtual multiplex social system by means of a generalised K-core. PLoS ONE **9**(12), e112606 (2014)
8. Mucha, P.J., Richardson, T., Macon, K., Porter, M.A., Onnela, J.P.: Community structure in time-dependent, multiscale, and multiplex networks. Science **328**, 876–878 (2010)
9. Szell, M., Thurner, S.: Measuring social dynamics in a massive multiplayer online game. Soc. Netw. **39**, 313–329 (2010)
10. Nicosia, V., Bianconi, G., Latora, V., Barthelemy, M.: Growing multiplex networks. Phys. Rev. Lett. **111**, 058701 (2013)
11. Wasserman, S., Faust, K.: Social Network Analysis. Cambridge University Press, Cambridge/New York (1994)
12. Seidman, S.B.: Network structure and minimum degree. Soc. Netw. **5**, 269–287 (1983)
13. Bollobás, B.: The evolution of sparse graphs. In: Graph Theory and Combinatorics, Proc Cambridge Combinatorial Conf in Honor to Paul Erdös, pp. 35–57. Academic, London (1984)

14. Dorogovtsev, S.N., Goltsev, A.V., Mendes, J.F.F.: k-core organization of complex networks. Phys. Rev. Lett. **96**, 040601 (2006)
15. Szell, M., Lambiotte, R., Thurner, S.: Multirelational organization of large-scale social networks in an online world. Proc. Natl. Acad. Sci. **107**, 13636–13641 (2010)
16. Castronova, E.: Synthetic Worlds: The Business and Culture of Online Games. University of Chicago Press, Chicago (2005)
17. Szell, M., Thurner, S.: Social dynamics in a large-scale online game. Adv. Complex Syst. **15**, 1250064 (2012)
18. Szell, M., Sinatra, R., Petri, G., Thurner, S., Latora, V.: Understanding mobility in a social Petri dish. Sci. Rep. **2**, 457 (2012)
19. Thurner, S., Szell, M., Sinatra, R.: Emergence of good conduct, scaling and zipf laws in human behavioral sequences in an online world. PLoS ONE **7**, e29796 (2012)
20. Szell, M., Thurner, S.: How women organise social networks different from men: gender-specific behavior in large-scale social networks. Sci. Rep. **3**, 1214 (2013)
21. Fuchs, B., Thurner, S.: Behavioral and network origins of wealth inequality: insights from a virtual world. PLoS ONE **9**(8), e103503 (2014). doi:10.1371/journal.pone.0103503
22. Bianconi, G.: Statistical mechanics of multiplex networks: entropy and overlap. Phys. Rev. E **87**, 062806 (2013)
23. Colomer-de-Simón, P., Serrano, M.Á., Beiró, M.G., Alvarez-Hamelin, J.I., Boguñá, M.: Deciphering the global organization of clustering in real complex networks. Sci. Rep. **3**, 2517 (2013). doi:10.1038/srep02517
24. Rapoport, A.: Spread of information through a population with socio-structural bias: I. Assumption of transitivity. Bull. Math. Biol. **15**, 523–533 (1953)
25. Granovetter, M.: The strength of weak ties. Am. J. Sociol. **78**, 1360–1380 (1973)
26. Davidsen, J., Ebel, H., Bornholdt, S.: Emergence of a small world from local interactions: modeling acquaintance networks. Phys. Rev. Lett. **88**, 128701 (2002)
27. Klimek, P., Thurner, S.: Triadic closure dynamics drives scaling laws in social multiplex networks. New J. Phys. **15**, 063008 (2013)
28. Harary, F., Ross, I.C.: A procedure for clique detection using the group matrix. Sociometry **20**, 205–215 (1957)
29. Colizza, V., Flammini, A., Serrano, M.A., Vespignani, A.: Detecting rich-club ordering in complex networks. Nat. Phys. **2**, 110–115 (2006)
30. Corominas-Murtra, B., Valverde, S., Rodríguez-Caso, C., Solé, R.V.: K-scaffold subgraphs of complex networks. EPL (Europhys. Lett.) **77**, 18004 (2007)
31. Corominas-Murtra, B., Mendes, J.F.F., Solé, R.V.: Nested subgraphs of complex networks. J. Phys. A: Math. Theor. **41**, 385003 (2008)

Chapter 11
Interbank Markets and Multiplex Networks: Centrality Measures and Statistical Null Models

Leonardo Bargigli, Giovanni di Iasio, Luigi Infante, Fabrizio Lillo, and Federico Pierobon

Abstract The interbank market is considered one of the most important channels of financial contagion. Its network representation, where banks and claims/obligations are represented by nodes and links (respectively), has received a lot of attention in the recent theoretical and empirical literature, for assessing systemic risk and identifying systemically important financial institutions. Different types of links, for example in terms of maturity and collateralization of the claim/obligation, can be established between financial institutions. Therefore a natural representation of the interbank structure which takes into account more features of the market, is a multiplex, where each layer is associated with a type of link. In this paper we review the empirical structure of the multiplex and the theoretical consequences of this representation. We also investigate the betweenness and eigenvector centrality of a bank in the network, comparing its centrality properties across different layers and with Maximum Entropy null models.

11.1 Introduction

Partly as a consequence of the crisis burst after the Lehman event in late 2008, financial networks are gaining popularity across policymakers, regulators and academics interested in systemic risk analysis. Networks are useful and natural tools

L. Bargigli (✉)
Dipartimento di Scienze per l'Economia e l'Impresa, Università di Firenze, Firenze, Italy
e-mail: leonardo.bargigli@unifi.it

G. di Iasio • L. Infante
Directorate General for Economics, Statistics and Research Bank of Italy, Rome, Italy
e-mail: giovanni.diiasio@bancaditalia.it; luigi.infante@bancaditalia.it

F. Lillo
Scuola Normale Superiore, piazza dei Cavalieri 7, Pisa, Italy
e-mail: fabrizio.lillo@sns.it

F. Pierobon
European Central Bank, Frankfurt am Main, Germany
e-mail: federico.pierobon@ecb.europa.eu

© Springer International Publishing Switzerland 2016
A. Garas (ed.), *Interconnected Networks*, Understanding Complex Systems,
DOI 10.1007/978-3-319-23947-7_11

to identify critical financial institutions as well as to understand how distress could propagate within the financial system.[1] The type of financial network considered in this paper is the one of direct exposures between financial institutions, which include credit relations, derivatives transactions, cross-ownerships, etc.

Interlinkages between any two financial institutions are more complex than the information that can be summarized in a single number (the weight of the link) and a direction, such as in a directed and weighted network. This is due to the fact that between two institutions there exists a multiplicity of linkages, each of them related to one class of claims/obligations. The interplay between different types of relations could be relevant for systemic risk analysis. In the network jargon, such a situation is best modeled with a *multiplex network* or simply *multiplex*. A multiplex is made up of several "layers", each of them composed by all relations of the same type and modeled with a simple (possibly weighted and directed) network. Since the nodes in each layer are the same, the multiplex can be visualized as a stack of networks or equivalently by a network where different types of links can coexist between two nodes, each type corresponding to a layer. Figure 11.1 shows a stylized representation of the interbank multiplex, where three layers are explicitly shown, together with the aggregated interbank network.

The study of financial multiplex is still at its infancy [5, 29]. This chapter builds on the analysis performed recently by the authors on the Italian Interbank Market (IIN) (see Ref. [5]). Taking advantage of a unique dataset collected by Banca d'Italia, Bargigli et al. [5] analyze in detail the evolution of the multiplex during the crisis (2008–2012).[2]

In this chapter, after a brief description of the Bargigli et al. [5] main findings on the multiplex representation of the interbank market, we add on a novel result by investigating an important network property, previously not considered, namely the network centrality of the banks. Node centrality is critical to identify the most important nodes in the network architecture. There are several different definitions of centrality, depending on the meaning given to the word "important" above. Here we shall consider betweenness centrality and eigenvector centrality, two widespread centrality metrics. Our main research questions can be summarized as follows: (i) Given a centrality measure of a node, how much is it correlated with the degree and the strength of the node? Is this correlation layer specific? (ii) Is the centrality (absolute or rank) of a bank approximately the same in each layer? Are there banks which are central in some layers but essentially peripheral in the others? (iii) How much can the centrality of a node be explained by a statistical null model that

---

[1]See for instance [1, 2, 6, 7, 9, 11–13, 18, 19, 21–24, 27, 31]

[2]Bargigli et al. [5] compare the topological properties of each layer, study the similarity between layers, i.e. how much one can learn from a layer knowing another one, and use a Maximum Entropy approach to investigate which high order topological properties of each layer can be explained by the statistical properties of degree and strength. In other words, the paper investigates how much the huge heterogeneity in degree and strength observed in interbank markets significantly constrains (e.g. for the assortativity) or is a sufficient statistics (e.g. for the core-periphery structure) other topological properties studied in the recent interbank network literature.

Fig. 11.1 Stylized
representation of the
multiplex structure of the
interbank market. Each node
is a bank, and links represent
credit relations. A layer is the
set of all credit relations of
the same type. The network in
red is the total interbank
market, obtained by
aggregating all the layers

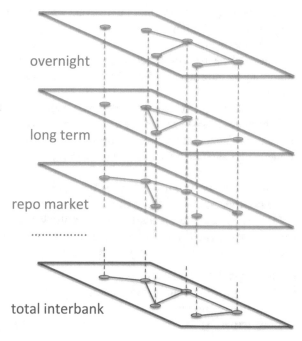

overnight

long term

repo market

. . . .

total interbank

preserves nodes' degree and strength? Is the centrality ranking mostly determined
by the degree (or strength) ranking?

The chapter is organized as follows. In Sect. 11.2 we present the dataset and
we introduce the construction of our multiplex. In Sect. 11.3 we review the main
findings of Bargigli et al. [5] about the Italian Interbank Network. Section 11.4
introduces several types of centrality metrics and presents our results, both on the
investigation of the centrality property of a bank in each layer and on the comparison
with Maximum Entropy null models. Section 11.5 concludes.

11.2 Data Description

We investigate a unique database of interbank transactions based on the supervisory
reports transmitted to Banca d'Italia by all institutions operating in Italy (see Ref. [5]
for details). We focus only on domestic links, i.e. transactions between Italian banks.
Information refers to end-of-year outstanding balances at the end of 2008, 2009,
2010, 2011, and 2012. This was a particularly delicate period for the Italian banking
system because of the concurrent sovereign debt crisis hardly hitting the Euro zone
and Italy in particular. Most banks operate in Italy through a large set of subsidiaries.
Distinguishing between intragroup and intergroup transactions is therefore crucial.
Since interbank lending and borrowing decisions are normally taken at the parent

company level, we assume that the relevant economic agents of the network lie at the group level. We thus focus on data on intergroup transactions consolidated at the group level. As shown in [5] this consolidation has a dramatic effect on the volumes and the network structure. In fact, in the investigated dataset and period the intragroup lending accounts for a fraction of the total volume ranging between 78 % and 89 %.

We represent the interbank market as a weighted and directed network, where a set of nodes (banks) are linked to each other through different types of financial instruments (edges). The direction of the link goes from the bank i having a claim to the bank j, and the weight is the amount (in millions of Euros) of liabilities of j towards i.

One specificity of our database is the availability of other information on the interbank transactions, namely the maturity and the presence of a collateral. As in [5] we use this information to build a multiplex representation of the interbank network, where each layer describes the network of a specific type of credit relation. Specifically, in terms of maturity we consider

- Overnight (OVN) transactions
- Short term (ST) transactions, namely those with maturity up to 12 months excluding overnight
- Long term (LT) transactions, namely those with maturity of more that 12 months

Considering collateralization we distinguish

- Unsecured (U) loans, i.e. without collateral
- Secured (S) loans, i.e. with collateral

Thus, for example, the symbol U ST stands for unsecured short term contracts. We have in total five possible combinations, since there are no secured overnight transactions. It is important to stress that our data on secured transactions only refer to OTC contracts, while secured transactions taking place on regulated markets and centrally cleared are not included in our database.

11.3   The Multiplex Structure of the Italian Interbank Market

In Ref. [5] Bargigli et al. presented one of the first in-depth analyses of the multiplex structure of an interbank market, investigating in particular the Italian market. Here we summarize our main findings. The first important observation is that the interbank market is dominated (in volume) by intragroup lending. To give an idea, in 2012 almost 78 % of the volume of the Italian interbank market was traded between two banks belonging to the same group, while only 22 % was intergroup lending. Since intergroup lending is the main channel of systemic risk propagation, in Ref. [5] and here we aggregate banks belonging to the same group and consider the network of banking groups.

When we consider the volume in the different layers, we observed that in 2012 the layer with the largest activity was *U LT* with roughly 41 %, followed by OVN with 28 % and *U ST* with 18 %. The two layers describing secured transactions have small volumes, namely 9 % and 2 % in the *S ST* and *S LT*, respectively. This is mostly due to the fact that today the majority of collateralized trades are operated through Central Counterparties (CCPs) and these trades are not included in our dataset. We found that the three unsecured layers are composed by many nodes, while the two secured ones are much smaller. In 2012 the Italian multiplex had 533 banks, and almost all of them were active in the three unsecured layers. In all layers there is a strong correlation between the degree of a node (i.e. the number of counterparts) and its strength (the total volume lent or borrowed). The Spearman correlation between these two quantities ranges between 0.49 and 0.94. Moreover degree and strength are also very correlated (>0.5) with bank size (as measured by the total asset). This means that the size of a bank is an important determinant of the amount traded in the interbank market and of the number of counterparts.

In Ref. [5] we also compared the topology of each layer individually. All the layers display a scale free property, i.e. the degree distribution has a power law tail characterized by a tail exponent between 1.8 and 3.5. This indicates a very strong heterogeneity of the system, with few important hubs and many low-degree nodes. Each layer has a small diameter, it is strongly reciprocal, and has a large clustering coefficient. Finally, it presents a disassortative mixing and has a clear core periphery structure. Below we will discuss the relevance of these two related last findings.

The multiplex structure of the system raises immediately the question of the similarity between layers, not in terms of their generic statistical properties, but on a link-by-link basis. In other words, in Ref. [5] we investigated how much the presence of a link between two banks in a layer is predictive of the presence of a link between the same two banks in the other layers. Interestingly, the answer is that layers are quite different one from each other, and the knowledge of the links in one layer gives little information about the presence of links in the other layers. On average across pairs of layers, the similarity, as measured by the Jaccard similarity, was roughly 17 % in 2012. This means that by observing a link in a layer, one can predict that the link exists in another layer only in 17 % of the cases. This low level of similarity, combined with the high level of similarity between the *same* layer in consecutive years (roughly 70 %), indicates that banks diversify their counterparts across different layers, but maintain stable relationships with the same counterparts in a given layer across the years. From a systemic risk point of view, the dissimilarity between layers tells us that the propagation of contagion can be significantly faster when one considers the multiplex structure, as compared with the contagion in individual layers. In fact, simple diffusion models on multiplex networks [20] show that diffusion can be very fast when the layers are "orthogonal" one another. In the second part of this paper we will consider a related question, namely how much the centrality measures of a node are specific of a layer or whether different nodes are the most central in different layers.

Finally, Bargigli et al. [5] compare the properties of the real multiplex network with those obtained under statistical null models, specifically with Maximum Entropy models (see Sect. 11.4 below). The Maximum Entropy principle allows to build explicitly probability distributions of graphs, which are maximally random and satisfy some given constraints. In particular we constrained the ensembles to have, on average, the same degree and/or strength sequence as the real network. This choice corresponds to the (weighted) configuration model. The aim of this analysis is to discriminate which high order properties of the real interbank networks are a mere consequence of the heterogeneity of degrees (or strengths) and which are instead genuinely new properties. We focused on disassortativity, reciprocity, and the presence of triadic motifs and we discussed which of these properties can be reproduced by the null model. To give a specific, yet important, example of the application of Maximum Entropy null models to interbank networks, we discuss briefly the case of the core-periphery structure.

Recently there has been an increasing interest toward the core-periphery structure of the interbank network [14, 17]. The core is defined as a subset of nodes which are maximally connected with other core members, while the periphery is the complementary subset made of nodes with no reciprocal connections but only with connections to the core [10]. Recently, Craig and von Peter [14] defined a tiering model in which core members without links with the periphery are penalized. Despite the fact that the two definitions of core-periphery are different, because of the different objective function, in Ref. [5] we found that the two models are highly correlated, i.e. the identified cores are essentially the same. The key question now is whether core-periphery is a genuine property of the interbank network or if it can be explained by the strong heterogeneity of degree. To this end in Ref. [5] we generated random samples from the Maximum Entropy ensemble where we fix either the average degree or the average weight of each node. By comparing the core-periphery subdivision in the real network or in the random samples, we conclude that they are very similar to each other.[3] Thus core-periphery subdivision (at least by using the aforementioned definitions) is a consequence of the large heterogeneity of degree. Since, as mentioned above, degree and bank size are strongly correlated, we conclude that in great part core-periphery structure is a consequence of the existence of large and small banks.

11.4 Centrality Measures

Centrality is a key concept of network theory, originally developed in social network analysis and rapidly developed to other types of networks. Broadly speaking, the centrality of a node (or of an edge) of a network is a measure of its importance,

---

[3]It is worth noticing that this can also be explained by using the result of Ref. [26], showing analytically that the subdivision in core and periphery according to the definition of Borgatti and Everett [10] is entirely determined by the degree sequence.

measuring, for example, how influential is a person in a social network, how critical is an element in an infrastructure network, what is the disease spreading capacity of an individual, etc. Despite being an important concept, the loose definition given above leads to several different proposed centrality measures, each of them able to capture some specific aspects of the concept of centrality. The review of all the centrality measures is beyond the scope of this chapter. Below we will discuss three measures of node centrality we are going to use in the analysis of the different layers of the IIN.

Degree centrality   Degree of a node is an obvious measure of centrality. A large number of links, in fact, is a symptom of the fact that the node is important for the connection of all the nodes which are linked to it. It is a local measure, i.e. it does not take into account the whole network, but only the local neighborhood of the node. Thus, while for small networks degree is a sensible centrality measure, for large networks it can miss important global characteristics of the importance of the node.

Betweenness centrality   One of the most popular global centrality measures is the betweenness centrality (often shortened as betweenness hereafter). It quantifies how frequently a node acts as a bridge along the shortest path between two other nodes. More formally, betweenness centrality of a node v is computed in the following way: for each pair of vertices (i, j) one identifies the N_{ij} shortest paths between them and computes the number $N_{ij}(v)$ of them that pass through v. The betweenness of node v is

$$C_B(v) = \frac{1}{(n-1)(n-2)} \sum_{\substack{i,j \neq v}} \frac{N_{ij}(v)}{N_{ij}}, \tag{11.1}$$

i.e. the average fraction of shortest paths passing through v, where the average is taken across all the pairs of vertices. The number of nodes in the network is n and the normalization factor in Eq. (11.1), and used in this paper, holds for directed graphs. Intuitively $C_B(v)$ measures how frequently a shortest path between two nodes passes through a given node.

Eigenvector centrality   Eigenvector centrality is defined in terms of the adjacency matrix $A = \{a_{ij}\}$, where a_{ij} can be either a binary or a non negative real value (weighted matrix). The vector $\mathbf{x} = (x_1, \ldots, x_n)'$, containing the eigenvector centrality x_i of node i, satisfies

$$A\mathbf{x} = \lambda^{max}\mathbf{x}, \tag{11.2}$$

i.e. it is a right eigenvector of the adjacency matrix corresponding to the maximal eigenvalue λ^{max}. Interpreting the network as a representation of a Markov chain with n states and transition probabilities proportional to the weights of the links connecting two nodes (states), it can be seen that the vector $\mathbf{x}$ is the stationary probability distribution of the chain. Interestingly the eigenvector centrality is related to Google's PageRank algorithm. In the context of systemic risk, it has been recently modified to define DebtRank [8] for measuring the centrality of the network of mutual exposures, and BankRank [15] for measuring the centrality of the bipartite network of banks and assets. In the present paper we will use the original definition of eigenvector centrality of Eq. (11.2).

Centrality Measures in the Interbank Multiplex

The aim of this subsection is to compare the centrality measures in different layers of the interbank network. The comparison of the degree of a bank in different layers has been investigated in depth in Ref. [5], also with respect to null models that preserve the degree in average (configuration model). For this reason in the paper we will consider mostly betweenness and eigenvector centrality.

As a preliminary analysis, Fig. 11.2 shows the distribution of betweenness in two important layers, overnight and unsecured medium term in 2012 (similar results are obtained for the other investigated years). The distribution is well fitted by a power law function with a tail exponent between 1.5 and 2. This fat tail behavior shows that there is a large heterogeneity of the betweenness centrality among the banks in all layers. In part this heterogeneity in the value of the betweenness is due to the scale free behavior of the layers. In fact, as shown in Ref. [5], the degree distribution of each layer is well Fit by a power law tail and the estimated values of the tail exponent are remarkably stable across layers and over time, ranging in [1.8, 3.5] and mostly concentrated around 2.3.

The relation between degree and betweenness of the banks in the overnight of the Italian Interbank Networks is shown in Fig. 11.3 (top panels), while the bottom panels show the relation between strength and betweenness. To have a numerical

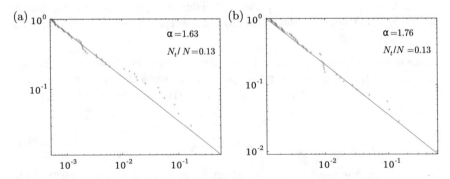

Fig. 11.2 Complementary cumulative distribution function of betweenness in (**a**) the unsecured overnight layer, and (**b**) in the unsecured long term layer in 2012. The plot is in double logarithmic scale and the *green line* is the best fit with a power law function. The tail exponent α is reported in the panels

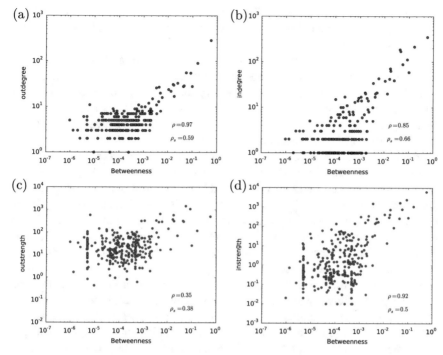

Fig. 11.3 Scatter plot of the betweenness and different nodes properties, namely (**a**) out-degree, (**b**) in-degree, (**c**) out-strength, and (**d**) in-strength. Each point is a bank group. The values of ρ and ρ_s are the Pearson and the Spearman correlation, respectively. Data refer to the unsecured overnight layer in 2012

value to quantify the dependence between the two variables, in each figure we report both the value of the Pearson correlation and the value of the Spearman rank correlation. It is worth noticing that in general the former is significantly larger than the latter. This is due to the strong correlation in the right tail of the two variables (i.e. for large and very connected banks). The top panels show that banks with large in- (or out-) degree are also the nodes with high betweenness centrality. The same holds when one considers the in- or out-strength (bottom panels), even if in this case this strong relation is evident only for a smaller interval of large values of strength.

This analysis shows that large (and/or more interconnected) banks are also typically those more central (as measured by the betweenness).[4] Thus, at least in the Italian Interbank Market, it is hard to discriminate between too big to fail and too interconnected to fail. Considering medium and small sized banks, the relation between centrality and degree and especially size becomes significantly noisier, and by looking at Fig. 11.3, it is possible to identify banks with moderate

---

[4]As mentioned in Sect. 11.3, Ref. [5] shows that nodes' properties (degree and strength) turn out to be correlated with balance sheet data of banks, in particular with the total assets.

strength but significant centrality. These institutions are likely playing a central role in intermediation in the considered layer or connecting subsets of banks which are only weakly connected.

We then investigate whether banks which are central in a given layer are also central in other layers. This is important because gives insight on the degree of specialization of some banks as intermediary for some type of credit. To answer this question we compute centrality measure in different layers and compare the values (or the ranking). We remind that the number of banks active in the different layers is different [5]. For example, in 2012, out of the 533 banks active in the interbank market, 532 had credit relations in the overnight market, 521 were active in the unsecured short term, 450 in the unsecured long term, and 45 in the secured short term. In order to compare layers with different number of nodes, we computed the centrality measures in the whole layer (i.e. including all the banks active in the layer), but we compared the centrality measures only within the subset of banks which were active in both layers. Therefore, when considering rankings, it should be kept in mind that these are not absolute rankings, but only rankings among the banks considered in the intersection.

Figure 11.4 shows the scatter plot of the betweenness in the overnight market either versus the betweenness in the unsecured long term (left panel) or versus the betweenness in the unsecured long term (right panel) market. In the figure we also report the value of the Pearson and Spearman correlation and, as explained above, the former is typically much higher of the latter due to the high correlation on the top left part of the scatter plot.

Comparing the short term and overnight layers (left panel), it is evident that top central nodes in one layer are also typically central in the other layer. For medium and low centrality, the correlation is much weaker. On the contrary, the right panel shows that, with the exception of three banks on the top right corner of the figure, there is a very weak correlation between the centrality in the overnight network and

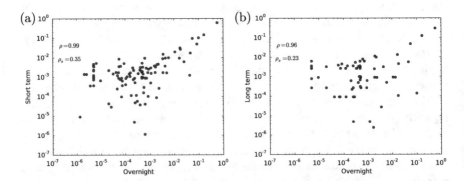

Fig. 11.4 Scatter plot of (**a**) the betweenness in the overnight market versus that in the unsecured short term layer, and (**b**) the betweenness in the overnight versus that in the unsecured long term layer. The year investigated is 2012. The values of ρ and ρ_s are the Pearson and the Spearman correlation, respectively

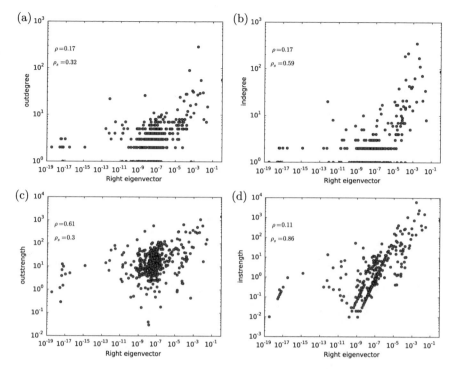

Fig. 11.5 Scatter plot of the eigenvector centrality and different nodes properties, namely (**a**) out-degree, (**b**) in-degree, (**c**) out-strength, and (**d**) in-strength. Each point is a bank group. The values of ρ and ρ_s are the Pearson and the Spearman correlation, respectively. Data refer to the unsecured overnight layer in 2012

in the long term network. This is an indication that in some cases, centrality of a bank, also with respect to the other banks, can be markedly different in different layers. Therefore central banks in the long term layer are not necessarily central in the overnight or in the short term layer.

Similar conclusions can be drawn by considering the eigenvector centrality. Figure 11.5 shows the scatter plot between the eigenvector centrality and in- and out-degree (top panels) and in- and out-strength (bottom panels). The correlations are significantly smaller when compared to those of the betweenness centrality (see Fig. 11.3). In this sense eigenvector centrality (and probably DebtRank) brings information on the importance of a bank which is not already contained in the basic measures of degree and strength.

Finally, in Fig. 11.6 we show the three dimensional plots of the eigenvector centrality in the three unsecured layers, namely overnight, short term, and long term. We show the results for the 4 years because we find a different behavior in the different years. In all years we find a very large dispersion of points, indicating that the eigenvector centrality in the three considered layers is very different. Thus

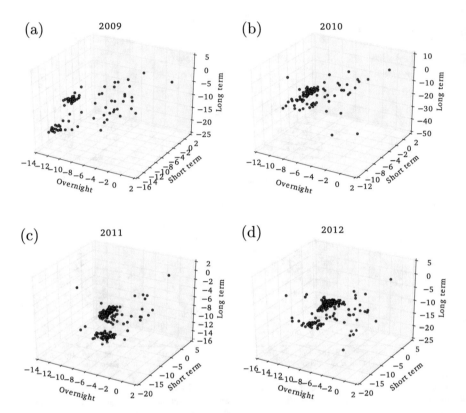

Fig. 11.6 Three dimensional scatter plot of the decimal logarithm of the eigenvector centrality in the three unsecured layers in the period 2009–2012. (**a**) 2009. (**b**) 2010. (**c**) 2011. (**d**) 2012

the importance of a node, as measured by the eigenvector centrality, is typically quite layer specific.

Comparison with Null Models

As a last analysis, we compare our findings on the centrality metrics with suitable statistical null models. In the last two decades complex network theory has introduced a large number of metrics able to capture many interesting aspects of the organization of networks, such as clustering, assortativity, core-periphery structure, etc. However, from a statistical point of view, it is not always clear which of these properties carries information not already contained, in simpler properties of the network. For example, one could ask whether the core-periphery organization observed in many networks (included the interbank ones) is a mere consequence of the large degree heterogeneity of the considered network. In fact, in a network with heterogeneous degree distribution, a core periphery emerges even if the links among nodes are assigned randomly.

To properly answer to these questions one needs to build statistical models of networks, which allow computing a probability distribution of graphs. Such a distribution is calibrated on the investigated real network, and it is chosen in such a way to preserve some low order properties (e.g. the degree of each node). Then one computes, analytically or computationally, the distribution of the considered high order property (e.g. the size of the core) and extracts from it a p-value for the statistics observed in the real data.

One of the most common methods for building null models is by using the Maximum Entropy Principle.[5] One looks for the probability distribution of graphs, $P(G)$, which maximizes the Shannon entropy

$$S[P(G)] = -\sum_{G} P(G) \ln P(G) \tag{11.3}$$

under suitable constraints, including the normalization $\sum_{G} P(G) = 1$. More details on the construction and estimation of Maximum Entropy models for networks are reviewed in Appendix 4 of Ref. [5].

Here we consider the so called Directed Binary Configuration Model (DBCM) (see Ref. [5]). In this case we impose $2n$ constraints, namely the average in- and out-degree of each node. This model can be fitted from a real network using Maximum Likelihood and one can then sample from a graph distribution of a DBCM calibrated on the a real network. We can compute the distribution of the betweenness centrality for each node of the network under the DBCM. We then compute the p-value of the betweenness centrality of each node observed in the real network.

Figure 11.7 shows the scatter plot of the betweenness centrality in the real network versus the one simulated in the DBCM. Firstly we note that there is a significantly high Pearson and Spearman correlation between the betweenness in the real and simulated network. We can therefore conclude that a significant fraction of the betweenness centrality property of a node is a consequence of the degree distribution. This evidence is in line with the empirical observation described above that betweenness centrality is very correlated with degree. However deviations from the null model can be identified by computing for each node the p-value, then comparing the centrality in the real network with the values obtained by a large sample of network realization. In Fig. 11.7 large dots correspond to nodes for which the (ME) null model is rejected with 1 % confidence. Interestingly, the nodes for which betweenness centrality is larger than expected from their degree (and the Maximum Entropy null model) are mostly large banks.

---

[5]See for instance [3, 4, 16, 25, 28, 30, 32].

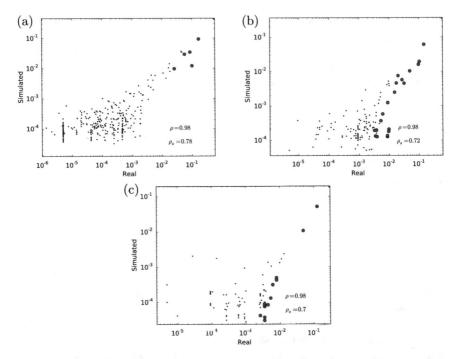

Fig. 11.7 Scatter plot of the betweenness centrality in the real network and in the one simulated from the Maximum Entropy (ME) ensemble corresponding to the directed configuration model, for (**a**) the overnight market, (**b**) the short term layer, and (**c**) the long term layer. Large dots correspond to nodes for which the (ME) null model is rejected with 1 % confidence. The values of ρ and ρ_s are the Pearson and the Spearman correlation, respectively. Data refer to 2012

11.5 Conclusions

This paper reviews the multiplex description of the interbank market proposed in Ref. [5]. In this description, each layer is a network of interbank exposures characterized by the maturity and the presence of a collateral. By using an unique dataset of the Italian Interbank market, we discuss the main findings related to (i) the similarity of the topological properties of the different layers, (ii) the link-by-link similarity of pairs of layers and (iii) the use of Maximum Entropy approach for the construction of layer-specific null models.

The main original contribution of the paper is the investigation of two centrality measures, namely betweenness and eigenvector centrality, in the multiplex describing the Italian Interbank market. We found that the correlation between the centrality of a bank in different layer is significant, but not extremely large. There are several medium-sized banks which are central in some layers and peripheral in others. Very large banks are typically central in all layers, but this can be due – at least partially – to the constraint given by their typically large degree. Interestingly, the centrality of large banks is not explained by Maximum Entropy null models preserving the degree of each node. This finding indicates deviations from centrality measure expected by the degree distribution. Finally, betweenness centrality is more correlated with degree than eigenvector centrality. This finding seems to indicate that the eigenvector centrality is less affected by information already contained in the degree, strength, and bank size.

Acknowledgements The views expressed in the article are those of the authors only and do not involve the responsibility of the Bank of Italy. This paper should not be reported as representing the views of the European Central Bank (ECB). The views expressed are those of the authors and do not necessarily reflect those of the ECB. The research leading to these results has received funding from the European Union, Seventh Framework Programme FP7/2007–2013 under grant agreement CRISIS-ICT-2011-288501 and from the INET-funded grant "New tools in Credit Network Modeling with Heterogeneous Agents". L. Infante contributed while visiting the NYU Department of Economics.

References

1. Abbassi, P., Co-Pierre, G., Gabrieli, S.: A Network View on Money Market Freezes, Document de Travail, 531, Banque de France (2014)
2. Allen, F., Gale, D.: Financial contagion. J. Pol. Econ. **108**, 1–33 (2000)
3. Bargigli, L.: Statistical ensembles for economic networks. J. Stat. Phys. **155**, 810–825 (2014)
4. Bargigli, L., Gallegati, M.: Random digraphs with given expected degree sequences: a model for economic networks. J. Econ. Behav. Organ. **78**, 396–411 (2011)
5. Bargigli, L., Di Iasio, G., Infante, L., Lillo, F., Pierobon, F.: The multiplex structure of interbank networks. Quant. Finance **15**(4), 673–691 (2015)
6. Battiston, S., Delli Gatti, D., Gallegati, M., Greenwald, B.C., Stiglitz, J.E.: Credit chains and bankruptcy propagation in production networks. J. Econ. Dyn. Control **31**, 2061–2084 (2007)
7. Battiston, S., Delli Gatti, D., Gallegati, M., Greenwald, B.C., Stiglitz, J.E.: Liaisons dangereuses: increasing connectivity, risk sharing, and systemic risk. J. Econ. Dyn. Control **36**, 1121–1141 (2012)
8. Battiston, S., Puliga, M., Kaushik, R., Tasca, P., Caldarelli, G.: DebtRank: too central to fail? financial networks, the FED and systemic risk. Sci. Rep. **2**, 541 (2012)
9. Bech, M.L., Atalay, E.: The topology of the federal funds market. Staff Reports 354, Federal Reserve Bank of New York (2008)
10. Borgatti, S.P., Everett, M.G.: Models of core/periphery structures. Soc. Netw. **21**, 375–395 (2000)
11. Boss, M., Elsinger, H., Summer, M., Thurner, S.: Network topology of the interbank market. Quant. Finance **4**, 677–684 (2004)
12. Caccioli, F., Catanach, T.A., Doyne Farmer, J.: Heterogeneity, correlation and financial contagion (2011). arXiv:1109.1213v1
13. Cont, R., Moussa, A., Santos, E.B.: Network structure and systemic risk in banking systems. In: Fouque, J.P., Langsam, J. (eds.) Handbook of Systemic Risk. Cambridge University Press, Cambridge (2013)
14. Craig, B.R., von Peter, G.: Interbank tiering and money center banks. Working Paper 0912, Federal Reserve Bank of Cleveland (2009)
15. Dehmamy, N., Buldyrev, S.V., Havlin, S., Stanley, H.E., Vodenska, I.: Classical mechanics of economic networks. http://arxiv.org/abs/1410.0104 (2014)
16. Fagiolo, G., Squartini, T., Garlaschelli, D.: Null models of economic networks: the case of the world trade web. J. Econ. Interact. Coord. **8**, 75–107 (2013)

17. Fricke, D., Lux, T.: Core-periphery structure in the overnight money market: evidence from the e-mid trading platform. Kiel Working Papers 1759, Kiel Institute for the World Economy, Mar 2012
18. Fricke, D., Lux, T.: On the distribution of links in the interbank network: evidence from the e-mid overnight money market. Kiel Working Papers 1819, Kiel Institute for the World Economy, Jan 2013
19. Gai, P., Kapadia, S.: Contagion in financial networks. Proc. R. Soc. A **466**, 2401–2423 (2010)
20. Gomez, S., Diaz-Guilera, A., Gomez-Gardenes, J., Perez-Vicente, C.J., Moreno, Y., Arenas, A.: Diffusion dynamics on multiplex networks. Phys. Rev. Lett. **110**, 028701 (2013)
21. Iazzetta, C., Manna, M.: The topology of the interbank market: developments in Italy since 1990. Temi di discussione, 711, Banca d'Italia (2009)
22. Iori, G., Jafarey, S., Padilla, F.G.: Systemic risk on the interbank market. J. Econ. Behav. Organ. **61**, 525–542 (2006)
23. Iori, G., De Masi, G., Precup, O.V., Gabbi, G., Caldarelli, G.: A network analysis of the Italian overnight money market. J. Econ. Dyn. Control **32**, 259–278 (2008)
24. Mastromatteo, I., Zarinelli, E., Marsili, M.: Reconstruction of financial networks for robust estimation of systemic risk. J. Stat. Mech.: Theory Exp. **3**, 11 (2012)
25. van Lelyveld, I., In 't Veld, D.: Finding the core: network structure in interbank markets. DNB Working Papers 348, Netherlands Central Bank, Research Department, July 2012
26. Lip, S.Z.W.: A fast algorithm for the discrete core/periphery bipartitioning Problem, Feb. 2011. arXiv:1102.5511v1
27. Martinez Jaramillo, S., Alexandrova-Kabadjova, B., Bravo-Benitez, B., Solorzano-Margain, J.P.: An empirical study of the Mexican banking system's network and its implication for systemic risk. Banco de Mexico, Working Paper, 2012-07 (2012)
28. Mistrulli, P.: Assessing financial contagion in the interbank market: maximum entropy versus observed interbank lending patterns. J. Bank. Finance **35**, 1114–1127 (2011)
29. Montagna, M., Kok, C.: Multi-layered interbank model for assessing systemic risk. Technical Report, Kiel Working Paper (2013)
30. Park, J., Newman, M.E.J.: Statistical mechanics of networks. Phys. Rev. E **70**, 066117 (2004)
31. Soramäki, K., Bech, M.L., Arnold, J., Glass, R.J., Beyeler, W.E.: The topology of interbank payment flows. Physica A: Stat. Mech. Appl. **379**, 317–333 (2007)
32. Squartini, T., Garlaschelli, D.: Analytical maximum-likelihood method to detect patterns in real networks. New J. Phys. **13**, 083001 (2011)

Chapter 12
The Financial System as a Nexus of Interconnected Networks

Stefano Battiston, Guido Caldarelli, and Marco D'Errico

Abstract In this Chapter, we describe the phenomenology of multilevel financial networks. Network analysis represents a useful tool for the analysis of financial systems, allowing, in particular, for a better understanding of the *mechanics* of systemic distress. However, the level of complexity reached by the financial system, coupled with the linkages arising to and from other economic sectors, calls for a more integrated approach that takes into account a whole series of networks. In this Chapter, we therefore describe the financial systems as a *nexus of interconnected networks*. By reviewing selected theoretical and empirical works and describing two methodological extensions for DebtRank, we show different arguments in favor of the adoption of a broader view of the network approach to finance.

12.1 Introduction

Systemic risk in finance denotes the risk of collapse of a major part of the financial market with the disruption of its critical functionalities. This notion can be declined in several ways. However, the main idea is that the collapse of an entire system (or a large part of it), as opposed to a single entity, can be triggered by specific interdependencies where the failure of a component (or a small number of components) propagates, and spreads the distress in a cascading process.

Network theory has recently emerged (from both a research and policy perspective) as one of the fundamental tools to quantify and assess the interconnected nature of social systems and therefore to model the spread of shocks within economic and financial systems. In-depth research has shown that financial organizations are

S. Battiston (✉) • M. D'Errico
Department of Banking and Finance, University of Zürich, Zürich, Switzerland
e-mail: stefano.battiston@uzh.ch; marco.derrico@uzh.ch

G. Caldarelli
IMT Alti Studi Lucca, Lucca, Italy

ISC-CNR, Rome, Italy

LIMS, London, UK
e-mail: guido.caldarelli@imtlucca.it

© Springer International Publishing Switzerland 2016
A. Garas (ed.), *Interconnected Networks*, Understanding Complex Systems,
DOI 10.1007/978-3-319-23947-7_12

particularly prone to be interconnected, through different types of relationships. Along the simple "size" of an institution, direct and indirect interlinkages (i.e. the network structure), represent another fundamental characteristic for the identification of systemically important institutions. The paradigm on which systemic importance is based must nowadays shift towards the inclusion of the "position" of an institution in the network.

Importantly, as pointed out in 2013 by the now Chair of the Federal Reserve Janet Yellen, although "some degree of interconnectedness is vital to the functioning of our financial system [...] Yet experience – most importantly, our recent financial crisis – as well as a growing body of academic research suggests that interconnections among financial intermediaries are not an unalloyed good".[1] These words were echoed, from the European side, by the President of the European Central Bank Mario Draghi, who pointed out not only that "the process of financial integration has created a myriad of complex linkages within the EU financial system" but also that "a more holistic view of interlinkages in the financial system is needed to understand how shocks are transmitted across the system and how to mitigate them".[2]

The first reason why linkages matter is that they can have ambiguous effects: on the one hand, they increase individual profitability while reducing the risk of the individual entity because of a more diversified structure; on the other hand, linkages allows for the propagation of contagion and distress, in certain cases with substantial amplification effects, thus increasing systemic risk. On this topic several issues remain open but much work has already been done in the recent years.

Therefore, *interconnectedness*, as a key structural property of financial systems, has posed the fundamental research question on whether the network of interdependences can induce higher levels of systemic risk, possibly amplifying small shocks. Network models have analyzed the *mechanics* of default contagion transmission [16, 23, 24, 36] to capture disruptions and systemic distress. However, they have been mostly limited to the banking sector (specifically, the interbank lending market) and to effects caused by the default of one or more banking institutions. Less attention has been devoted to the spillovers onto other macroeconomic sectors (including the real economy) and potential feedback effects that these exposures imply [18]. Moreover, as described later on, not only the propagation of defaults is important, but also the propagation of distress [9].

In this work, we argue that the financial system should be seen as a *nexus of interconnected networks*, rather than an autonomous and independent networked system where shocks originates from outside. More specifically, the key idea of this Chapter is to investigate a mechanics of shock transmission, starting from heteroge-

---

[1] Janet L. Yellen, *Interconnectedness and Systemic Risk: Lessons from the Financial Crisis and Policy Implications*. Speech at the American Economic Association/American Finance Association Joint Luncheon, San Diego, California, January 4, 2013. URL: http://www.federalreserve.gov/newsevents/speech/yellen20130104a.htm

[2] European Systemic Risk Board, *Hearing before the Committee on Economic and Monetary Affairs of the European Parliament*, Introductory statement by Mario Draghi, Chair of the ESRB, Brussels, 18 February 2013. Available at https://www.esrb.europa.eu/news/pr/2013/html/is130218. en.html.

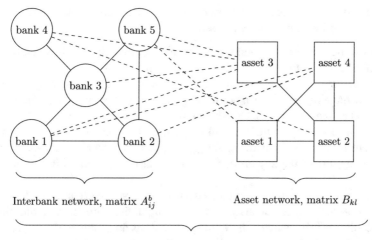

Asset network, matrix B_{kl}

Bipartite bank/asset holding network, matrix A^e_{ik}

Fig. 12.1 Representation of the financial system as a nexus of three interconnected networks: the interbank exposure network (matrix A_{ij}) and the asset (correlation) network are interconnected via the bipartite network of the asset holdings by banks

neous balance sheet exposures and incorporating the feedback relationship derived by common asset exposures. Consider, as the starting point for this discussion, Fig. 12.1, which reports a stylized network representation of one of the levels of the nexus. The interbank lending structure is on the left part of the figure, and is represented in Sect. 12.2 by matrix $\mathbf{A}^b$. Alongside this network, we can identify a network of assets, i.e. a network capturing structural dependencies between single asset classes. The dependencies can go beyond simple price correlations and include more "mechanical" dependencies such as those occurring between a derivative and its underlying. These two networks are further interconnected via a bipartite network of portfolio holdings (banks' exposure to external assets) for each bank.

In modern financial systems, banks indeed have a very heterogeneous and complex portfolio, which might include financial exposures of very different nature and levels of complexity: from simple mortgage loans and shares to very complex derivatives [10]. More to this, counterparties in the networks can be actors in different macroeconomic sectors and therefore feedback effects can arise also along this dimension.

Notwithstanding the increased awareness of both researchers and policymakers about the potential associated to modeling such systems via network theory, little work has been done in understanding how the nexus is shaped and works. We then envision the nexus as having to deal with two main characteristics:

1. it must take into account both the internal nature of the banking system (*within* the sector) and the external (*between* the sectors) connections with other institutional sectors.

2. it must take into account the structure of balance sheets, including feedback relationships with their external assets.

With respect to the first point, the financial sector is indeed linked with other parts of the economic system on different levels. In fact, the current configuration of the financial system seems to even favor spillovers, so that shocks may potentially propagate quite rapidly and trigger an amplification of the initial distress (however small). The financial crisis has shown that financial linkages among different actors in the economy (including banking institutions, non-banking institutions and firms) are a clear mechanism for the propagation of financial distress, therefore representing one of the key elements of potential financial instability or systemic risk.

On the above basis, we also produced one of the first ways to model interconnections also *outside* the financial system. In this light, the main research question we want to address is to evaluate the validity of analysis done only in the intra-sectoral networks. Indeed, macroeconomic literature considers aggregated economic sectors as separated "blocks", therefore neglecting the *multiplier effect* arising from the interconnections within each block. However, network effects are relevant also within an individual sector: considering macroeconomic blocks as mere aggregates would then lead to neglect the multiplier effect within the block. On the other hand, the behavior of a single macroeconomic block cannot be assessed outside of its interconnections with other macroeconomic blocks.

As regards the second point, financial institutions are interconnected not only through different types of relationships, maturity, and instrument types, but the linkages also arise from common exposure in terms of asset classes. This feature may very well amplify an initial shock via *fire sales*. Therefore, shocks can be transmitted back and forth via these common exposures. Note that such fire sales cannot be modeled if only one agent is considered. Shocks on the assets of a bank trigger sales in order to restore capital requirements that then spillover onto other financial institutions along two channels: (1) the interbank lending exposures and (2) the reduction in price of assets held by other banks due to fire selling.

However, these two channels are further intertwined as sudden variations in the value of assets held by a bank can have direct and indirect spillovers on other economic sectors, and in particular the real economy (including non-financial corporations, households and government) therefore exacerbating the original shock.

12.1.1 Related Work

In the aftermath of the crisis, substantial academic and policy-oriented research has been carried out in order to better understand the nature and the potential implication of systemic risk. In particular, several works focus on the main idea that one of the fundamental roles in determining financial instabilities is played by financial interlinkages. In this light, network models can be useful tools for the mathematical formulation of systemic risk [29].

Within the network approach to systemic risk, several questions still remain unanswered. However, considerable work has already been done in the recent years (see, among other, Refs. [3, 4, 7, 8, 12, 19, 26–28, 30, 39]). Not limited to contagion, financial interlinkages are also relevant for liquidity provision [27].

There is also an endogeneity problem that has been explored: financial institutions have incentives to become too-connected-to-fail and/or too-correlated-to-fail [1], therefore forming strongly knit structures [15, 17, 20, 21, 32, 38, 41, 42] and gain exposures to similar risks [26]. As a consequence, the system as a whole structures itself such that the incentives of the individual financial institution in terms of risk taking have been altered, possibly allowing groups of institutions with high market power to influence the debate on regulation.

12.2   Multi-level Financial Networks and Leverage

The main idea of this Section is that the level of complexity described above is embedded within the balance sheet structure of financial institutions and that we can investigate the mechanics of shock transmission starting from this representation. In particular, within a stylized balance sheet model, the key quantity we consider is leverage. We argue that leverage plays a key role in quantitatively determining the mechanics of distress. Leverage is simply defined as the ratio between the total assets of a bank and its equity. Obviously, leverage varies in time as the size of the balance sheet but there is evidence that banks might target leverage to a specific value [2, 40] when facing shocks, often for regulatory reasons. The leverage ratio can be simply decomposed (*additively*) by exploiting its linear additivity. However, each of the contributions to leverage carry with itself a high degree of interconnection, that results into a leverage network, captured by a weighted adjacency leverage matrix.

We show, by means of simple calculations, that even though leverage can be decomposed additively by simply exploiting the linear relationship stemming from balance sheets, its effects in a network framework are of multiplicative nature, therefore leading to a compounding effect of initial exposures.

12.2.1   Leverage Decomposition

We reformulate the balance sheet model in terms of leverage and devote particular attention to its decomposition. The main point here is to show that leverage can be decomposed *additively* but compounds *multiplicatively* in the distress process, exacerbating potential *within* and *between* network effects. Consider the standard

Assets A_i	Liabilities L_i
$A^e_{i1} = l^e_{i1} E_i$	D^e_{i1}
$\vdots$	$\vdots$
$A^e_{ik} = l^e_{ik} E_i$	D^e_{ik}
$\vdots$	$\vdots$
$A^e_{iM} = l^e_{iM} E_i$	D^e_{iM}
$A^b_{i1} = l^b_{i1} E_i$	$D^b_{i1} = A_{1i} = l^b_{1i} E_1$
$\vdots$	$\vdots$
$A^b_{ij} = l^b_{ij} E_i$	$D^b_{ij} = A_{ji} = l^b_{ji} E_j$
$\vdots$	$\vdots$
$A^b_{iN} = l^b_{iN} E_i$	$D^b_{iN} = A_{Ni} = l^b_{Ni} E_N$
	E_i

Fig. 12.2 Expansion of bank i's balance sheet, showing the banking system as a nexus of interconnected networks

balance sheet identity for bank i:

$$A_i = D_i + E_i \tag{12.1}$$

where A_i represent the total assets, D_i the total liabilities and E_i the value of its equity. Equity, in this context, can be interpreted as the buffer for the node/bank within the network to suffer losses before default. Equation (12.1) reflects the schematic representation provided in Fig. 12.2. The *total leverage* of i is defined as the ratio between its assets and its equity:

$$l_i = \frac{A_i}{E_i}. \tag{12.2}$$

We divide asset and liabilities of i into two main categories (more detailed classifications can be easily incorporated in this representation): *internal* (with respect to the banking system, i.e. the interbank) and *external* (i.e. outside the banking system.). These sub-categories have a correspondence, respectively, to the previously described notions of *within* and *between*, that will be further explored in Sect. 12.3. A more granular subdivision is represented by the values of the asset and liabilities within each counterparty, i.e. the $1, \ldots, k, \ldots, M$ external assets (liabilities) and the $1, \ldots, j, \ldots, N$ interbank assets (liabilities).

The total leverage can be now decomposed as the sum of the leverages with respect to each external asset and each counterparty:

$$l_i = \frac{A_i}{E_i} = \frac{\overbrace{A_{i1}^e + \ldots + A_{ik}^e + \ldots A_{iM}^e}^{\text{external assets}} + \overbrace{A_{i1}^b + \ldots + A_{ij}^b + \ldots + A_{iN}^b}^{\text{interbank assets}}}{E_i}$$

$$= \underbrace{l_{i1}^e + \ldots + l_{ik}^e + \ldots + l_{iM}^e}_{\text{external leverage (between)}} + \underbrace{l_{i1}^b + \ldots + l_{ij}^b + \ldots + l_{iN}^b}_{\text{interbank leverage (within)}}. \qquad (12.3)$$

We can then expand the balance sheet identity by representing it in terms of leverage. Identity Fig. 12.2 can be then rewritten as:

$$\left(\sum_{k=1}^{M} l_{ik}^e + \sum_{j=1}^{N} l_{ij}^b - 1 \right) E_i = \sum_{k=1}^{M} D_i^e + \sum_{j=1}^{n} l_{ji} E_j$$

where l_{ji} represents, consistently with our notation, the contribution of the total leverage of j represented by the exposure towards i.

12.3   Networks of Networks: Effects *Within* and *Between*

Even though the banking system is now fully understood as a highly interconnected system along several dimensions, the recent crisis has unequivocally showed that significant shocks may originate from other macroeconomic sectors and spill over onto others. The ongoing problems in financing the real economy represent, for example, compelling evidence that crises in the banking system amplify shocks that then reverberate onto the real economy. For example, the European Commission, aware of the "the impact of external financing difficulties on the real economy", has pointed out that one of the key policy concerns relates to the steps to be taken in order to make sure that "banks resume their role as financiers of new business activities and lenders to viable firms, in particular in those parts of the corporate sector that rely mostly on bank funding" [25]. The relationship goes in the opposite direction as well, as the traditional funding activity of banks has been represented by deposits from households and firms. The fact that several banks had to be bailed out with complex rescue programmes is also a key example of this.

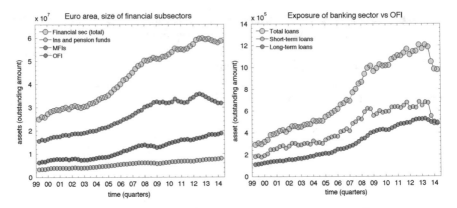

Fig. 12.3 (*Left*). Assets size of the financial sector its and composing sub-sectors (including Rest of the World, mil euro) from 1999 to 2014 (quarterly observation). (*Right*). Assets size of financial sector and sub-sectors (including Rest of the World, mil euro) from 1999 to 2014 (quarterly observation) (Source: ECB Statistical Data Warehouse)

To illustrate the particular structures we aim at describing, consider Fig. 12.3. The plot on the left represents the time evolution of the size (measured in terms of total assets, outstanding amount) of the financial sector in the euro area. It is clear that the staggering increasing occurred from 2001 to the recent years is mostly driven by MFI (monetary financial institutions, mostly banks) and other financial intermediaries (which include financial institutions of different nature). Insurance and pension funds showed a relatively smaller increase. The increase in assets can reflect higher level of risk (both individual and systemic), but this would not be of concern for the financial system as a whole if the different sub-sectors were not heavily exposed to one-another. It turns out that, on the contrary, these sectors are heavily interconnected, as shown in the right side figure, which reports the dynamics of the exposures (in terms of loans) from the banking sector towards other financial intermediaries. Linking the two figures, it is immediate to observe that the steady increase in total assets *within* the banking sector has corresponded to an increase in the interconnection towards another financial sector, therefore increasing the weight of the linkage *between* the two sectors.

The consequences of this are not trivial. Banks might be exposed to investment funds carrying higher degree of risk than expected: a shock for investment funds can reverberate, via the balance sheet structure, onto the banking system, then be amplified within the interbank lending system, and subsequently feed back onto other sectors. Furthermore, the figure also hints at another peculiarity of the interconnections occurring in the financial system: linkages can be of different nature and type. In fact, shocks in short terms loans (often due to liquidity problems) can lead to reduce exposure in long-terms loans (e.g. mortgages to household or long-term loans to non-financial companies).

Using disaggregated information on balance-sheet level is therefore crucial in order to understand potential systemic threats along the different interconnected networks. This intrinsic characteristic of financial systems can indeed induce higher levels of systemic risk, possibly amplifying small initial shocks. In spite of that, the state of the art in financial network modeling predominantly focuses on the understanding of particular features of systemic risk within the banking system, with a particular attention to the interbank lending market. This approach, although highly valuable for a deeper understanding of the direct and indirect interlinkages between banks, obviously leads to discarding the above mentioned interlinkages via exposures to external assets and sources of lending.

The picture we have so far sketched obviously needs a further enrichment, by considering also other sectors in the economy, including households, non-financial corporations and the government. The particular level of interdependence we aim at describing here is aptly captured by bilateral exposures between macroeconomic sectors. By analyzing such bilateral exposures, one can see how shocks originated in one sector might indeed engender spillovers onto other macroeconomic sectors, which in turn feed back to the banking systems itself. A country-based level analysis of this type can be found in Ref. [18].

The view we hereby adopt is, therefore, a combination of a *within* as opposed to a *between* approach. From this point of view, network effects *within* a sector (and particularly, the banking sector) are further exacerbated by network effects *within* the banking sector.

Given the potential systemic risk arising from this further levels of interconnection, the next generation of network-based systemic risk models must necessarily take into account these two levels. In this light, we aim at underlining the expanding the notion of interconnectedness in the financial system in order to take into account the above mentioned within and between nature of the financial system seen as a network of networks.

However, a bank's balance sheet is not only composed of interbank loans. Figures 12.4, 12.5 and 12.6 show the exposures of different macroeconomic sectors, including banks, in terms of total assets (outstanding amount, as a sum of deposit plus short and long terms loans) for the euro area in the first quarter of 2005, the third quarter of 2008 and the first quarter of 2014. This type of macroeconomic relationship is often referred to as "who-to-whom" and "provides an overview of the activities between the various sectors of the economy".[3] From a national accounting perspective, the main statistical source for these types of linkages is represented by flow-of-funds data, a fundamental concept in the system of national accounts that captures the interdependencies amongst the balance sheets of different institutional (macroeconomic) sector. A staggering increase in the total assets of banks is clearly visible from 2005 to 2008. An overall increase of the weight of the linkages can be also seen from the Figures.

Figure 12.6 (first quarter of 2014) shows how a shock in the external assets of a bank (red rectangle), such as mortgage-backed securities, can be exacerbated from

---

[3]Euro area accounts: SDW reports, available from the ECB Statistical Data Warehouse at http:// sdw.ecb.europa.eu/servlet/desis?node=1000002778

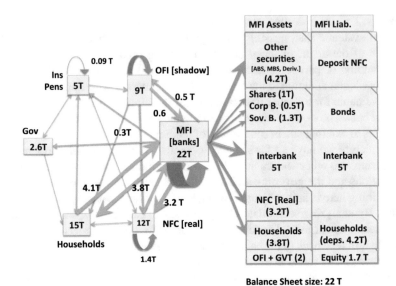

Fig. 12.4 The "Macro Network", 2005 first quarter. Assets (deposits, short and long-term loans)

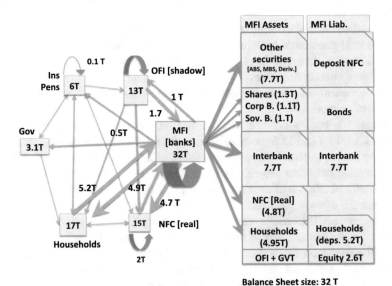

Fig. 12.5 The "Macro Network", 2008 third quarter. Assets (deposits, short and long-term loans)

the two networks: common assets and the interbank. Afterwards, this effect can spill over onto other economic sectors. From a modeling standpoint, this implies to develop techniques and tools that consider the aggregated macroeconomic blocks taking into account the multiplier effect of shocks within that and subsequently

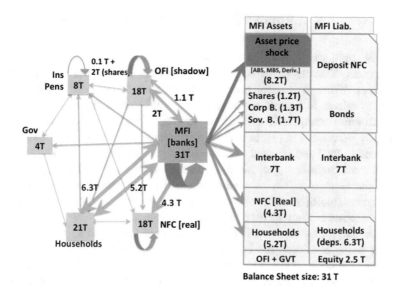

Fig. 12.6 The "Macro Network", 2014 first quarter. Assets (deposits, short and long-term loans)

spread towards other blocks. These further enriched frameworks should ideally be able to consider the coupling effect between different sub-sectors in the financial system itself (such as investment and pension funds, insurance corporations, money markets, etc.), the impact onto other sectors and possible feedback effects.

From a network perspective, a few important comments are in place, with reference to the left hand side of each figure. At first, one can observe households being exposed for a significant amount towards banks in terms of deposits. The typical maturity transformation banking activity takes place in terms of loans (short and long terms) towards households and non-financial corporation. However, the banking sector is exposed by a non-negligible amount towards other financial corporations (OFI), which partially include shadow-banking activity [5]. This macro network is coupled with a balance sheet expansion for the banking sector. A shock on the asset side of the balance sheet can therefore trigger a reverberation (with further amplifications) in the interbank lending market and subsequent spillovers onto the other institutional sectors.

This representation shows the need for a more granular representation of a bank's balance sheet in order to quantify the exposure towards other institutional sectors. In addition to this, more information on banks' balance sheets can be crucial in understanding also the level of exposures to common assets (see Sect. 12.6 for details on the modeling framework) and therefore to common shocks that reverberate from one layer to another.

12.4   Network of Financial Networks: Selected Models

In this Section, we review and put into our "nexus of networks" context some of the recent advances in financial network models that see the financial system as a series of interdependent networks. From a modeling perspective, the first work [36] relates to default cascades due to credit runs and fire sales that deals with different paradigmatic network topologies. We will then decline DebtRank (originally introduced in [9]) in a leverage framework. Last, we discuss the model introduced in [40] and proceed to discuss how procyclicality in banks exposed to external asset can affect systemic risk. On the empirical level, Roukny et al. [37] models the German interbank market where the interbank network is seen as a multi-layered (multiplex) network, whereas Puliga et al. [35] aims at reconstructing network exposures from the time series of CDS spreads.

A summary of these works is provided in Table 12.1, with an overview of the network levels considers, the contagion dynamics, the relevant quantities in the papers and the main results. This set of works provides with a series of compelling arguments towards the inclusion of several network layers into the understanding of systemic distress.

12.4.1   Default Cascades with Credit Runs

Theoretical Framework

The balance sheet structure of banks is, once again, the main determinant of a model originally proposed in Ref. [6]. The model features N banks, with the usual balance sheet structure: A_i^b being the interbank assets, A_i^e being the external assets (i.e. outside the banking system), L_i^b and L_i^e represent respectively the interbank and external liabilities. External assets are further classified as short-term (liquid) assets A_i^{es} and mid-long-term (less liquid) assets A_i^{em}. The level of liquidity of the mid-term assets will determine the loss a bank will face when selling them. Similarly to the asset side, liabilities can be classified in mid-long-term liabilities (both interbank and external), L_i^{em} and short-term liabilities L_i^{ES} owed to external creditors.

In line with previous well known work [23], a bank i in the banking network system defaults when its equity e_i becomes negative. The model investigates how market liquidity on the external assets, coupled with interlinkages arising from interbank exposures may influence the total number of defaulting banks in the system. Consistently with the approach used in this work, let:

$$\eta_i = \frac{A_i - L_i}{A_i^b} = \frac{E_i}{A_i^b}$$

which can be seen as an *interbank equity ratio*. Defaults for i will then occur when $\eta_i < 0$. When some of the borrowers of i face default, this will trigger the usual

Table 12.1 Recent works on systemic risk and networks

Theoretical works

Reference	Network levels	Contagion dynamics	Relevant quantities	Main results
Default cascades [36]	1. Interbank 2. short term external funds	1. Default and 2. fire sales	Default cascade size (phase diagram)	No optimal topology
DebtRank [9]	1. Interbank 2. [possibly] external assets	Distress and defaults	Systemic impact of individual or groups	"There's more to impact than balance sheet size"
Procyclicality	Interbank common assets	Distress, fire sales balance sheet management	Time to default	Counterproductive effects of capital requirements in presence of market illiquidity

Empirical works

Reference	Network levels	Type of data	Relevant quantities	Main results
Interbank (Germany) [37]	Interbank lending, derivatives (CDS)	Exposures (monthly 2002, 2012)	Comparison of network properties over time (CDS)	Most topological properties stable despite the financial crisis
CDS networks [35]	CDS (simulated) common asset	Time series (different network construction methods from comovements)	Network properties and Group DebtRank	Network effects are important. A combination of network effects with macro-economic indicators might be needed to capture the building-up of systemic risk

mechanism of losses on i. By considering the limiting case of zero recovery rate, creditors of i may opt not to renew short-term loans: i is then forced to sell assets in order to pay back those liabilities. Quite intuitively, bank i will first sell liquid assets and, only at a second moment, less liquid assets. If the market for the latter type of assets is rather liquid, then the bank might have to sell the assets for a lower price than market price, thus leading to so-called "fire-sellings".

Within this framework, $L_i^s - A_i^s$ represents the difference between the amount bank i must pay back and the amount it can liquidate in a sufficiently short time. Including fire sales on the less liquid assets, one can express the loss for bank i in terms of its interbank assets as follows:

$$\beta_i = (q-1)\frac{L_i^s - A_i^s}{A_i^b}.$$

implying $\beta_i = 0, \forall i$ when the market is perfectly liquid.

The equity of bank i can be reduced by two mechanisms taking place within this model:

1. *Counterparty.* When a counterparty j of i defaults, then i faces a loss of A_{ij}, without changing its liabilities;
2. *Credit run.* As a consequence, short-term external creditors might withdraw liquidity in case the number of defaults among i's counterparties reaches a certain level.

The second mechanism depends indeed on i's capital ratio. Let $k_{fi}(t)$ be the number of borrowers of i that have defaulted between time 0 and t, then the threshold for the credit run will be set in the following way:

$$\frac{k_{fi}(t)}{N} > \eta_i(0)/\gamma$$

γ is a parameter that captures the sensitivity of the external creditors with respect to bankruptcies for the borrower of i. This dynamics therefore creates a *cascade of defaults*. The *size* of the cascade is simply measured with the number of defaulted banks and computed at each time t of the above described mechanism. The process starts with an exogenously determined number of defaults and continues until no additional defaults are observed.

Interestingly for the scope of this work, from an aggregated perspective, the model can capture the macroscopic dynamics of the cascade of defaults depending on the initial conditions that have triggered the bank run. Furthermore, under some approximations concerning the topology of the network and the level of correlation of defaults within the neighborhood of every bank, the total size of the cascade can be computed analytically [7]. Confirmation of these results and further extensions to different network structure can be found in Ref. [36].

As the model centers upon the capital ratio η_i, it is then interesting to examine and interpret some results related to the relationship between capital ratios requirements and systemic risk. In particular, the model helps to understand the resilience of the system from a network perspective or, in other words, the impact of different topological structures to the global systemic risk. Even if the model is simple in its structure, conditioning to a specific topology, coupled with the various parameters, brings various degrees of freedom, which can be addressed in the following way:

1. Consistently with previous literature on networks (see Ref. [22]) and, in particular, financial networks (see Ref. [26]), the first modeling choice relates to the initial selection of the shocked banks, i.e. whether the initially distressed banks should be chosen at random (randomly chosen shock) or based on their degree centrality (targeted).
2. Different scenarios in the correlation between a bank's capital ratio and its degree are analyzed. When no correlation between capital ratios and degree is present, this implies that capital ratios are assigned independently from the total number of a contracts the bank holds. On the contrary, in case of positive correlation,

higher degrees correspond to higher capital ratios. The opposite holds in case of negative correlations.

3. Different degrees distributions are examined for three archetypal network topologies, namely: scale–free networks (power-law degree distribution), random graphs (Poisson degree distributions) and regular graphs (i.e. all nodes have the same degree).

4. Last, the parameter β varies to capture different levels of liquidity.

Different combinations of the above-described situations can lead to the analysis of different scenarios. The model shows interesting results in the way the cascade size can be related to different network measures such as the average out-degree (i.e. the average number of out-going exposures) and the initial average individual capital ratio m.

From a network-theoretical point of view, the networks considered are directed and weighted. The direction of a link follows the exposure from a lender to a borrower, and the value of the weight of the link is the outstanding amount. The model further assumes that banks lend equally to other banks they are exposed to, i.e. an amount $1/k_i$, $\forall j$, where k_i is agent i's number of borrowers. Out-degree then measures the number of borrowers, whereas the in-degree measures the number of lenders. As a benchmark case, the model investigates the case where in-degree and out-degree are correlated.

Experimental Set-up

The initial value of the capital ratio $\eta_i(0)$ for each bank is assigned randomly from a normal distribution with mean μ and variance $\sigma_\rho^2 = \sigma^2/k$ where μ and σ are exogenously given. In other terms, at time 0, capital ratios are drawn at random from a normal distribution $\eta \sim N(\mu, \sigma_\rho)$, therefore immediately implying the fraction of defaulted banks at the beginning (when $\eta_i(0) < 0$). In addition to the initially defaulted banks, a further exogenous shock is added in terms of initially defaulted banks (a fraction y_0).

A key assumption is made within this framework: a large number of credit counterparties implies a small variance in the return of the credit portfolio of each bank and reduced the variance of individual robustness. This is reflected in the relationship between σ_ρ and the average number of connections of the banking system k. For a more thorough discussion about the choice of the parameters, see Ref. [7]. All simulations in this setting are run for 1000 banks where, for each simulations, a network is generated and the cascading process takes place. Given a topology, for each pair of parameters (b, m) and (k, m), the mean and standard deviation of the cascade process is taken over the 1000 simulations. In this way, it is possible to determine a curve representing the so-called *frontier* (within the parameter space) that separates regions where either small or large cascades occur.

Empirical Evidence: The e-MID Interbank Market

An interesting analysis on empirical data is carried for daily observations of the Italian interbank money market network, provided by the Italian electronic Market for Inter-bank Deposits from January 1999 to December 2011,[4] aggregated on a monthly level. In this way, the model can be analyzed in an empirical context, where network evolution spans different periods (including the crisis).

Phase diagrams are provided in order to illustrate the so-called *systemic risk frontier*. High levels of systemic risk (i.e. very large-sized cascades, region in color) and low levels of systemic risk (small sized-cascades, white region) can be observed within the parameter space of market illiquidity β and average capital ratio m. Different colors are associated to different topologies and it can be observed what value of the average capital ratio would move the system towards a safe region.

Figure 12.7 captures the system in the case of positive/negative correlation between the capital ratios and the degree. The space for the scale-free topology lies below the other ones, implying that this type of topology is more robust against random shocks in the case of higher capital ratios for higher-degree banks (the opposite happens when higher-degree banks have lower capital ratios). Importantly, these differences are relevant only when market illiquidity is high enough ($b > 0.3$). When shocks are targeted to more central nodes, scale-free topology proves to be extremely fragile for any level of market illiquidity (see original paper for the results). This finding relates to previous works in network theory, where the notion of fragility of the network is measured in terms of the size of the largest connected component that survives a shock.

Figure 12.8 shows the frontiers for the period 1999–2008 (smooth and linear dependence on illiquidity β) which can be separated from those within the period 2009–2011. The latter curves lie below the others, showing that the average robustness is less sensitive to illiquidity. Lower systemic risk can be found for the year 2009 (post Lehman Brothers) whereas higher systemic risk is found for the frontier corresponding to the years 2007 and 2008. The post Lehman period is characterized by significant increases in the interbank rates coupled with takeovers by central banks in order to provide liquidity: during this period, banks withdrew from the interbank market, becoming less prone to lend to each other and trading more with central banks. This historical fact finds a correspondence to a smaller sensitivity to the illiquidity parameter with respect to previous years. Later, in 2010 and 2011, banks started to resort again to the interbank market for the provision of liquidity.

A particular historical moment can be found in 2009, with the guarantees by the European Central Bank (ECB). The ECB's provisions, coupled with a rise in the interbank rates, led to banks being less active in the interbank. As a consequence, illiquidity plays a lesser role in determining the overall cascade size,

---

[4]Data are maintained by e-MID S.p.A, Società Interbancaria per l'Automazione, Milan, Italy (e-MID).

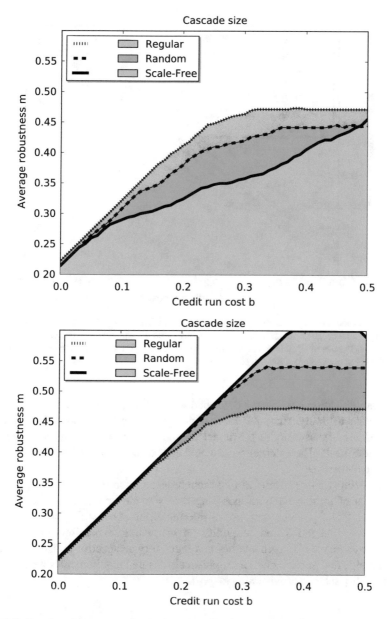

Fig. 12.7 Frontier of large cascades in the space (b, m) representing the average capital ratio across banks and market illiquidity. (*Upper figure*) Random exogenous defaults and positive correlation between degree and individual robustness. (*Lower figure*) Random exogenous defaults and negative correlation between degree and individual robustness, Networks have an average degree, $\bar{k} = 20$. Other parameters values are: $\gamma_s = \gamma 10^{-3}$, $\gamma_s = 0.13$ $\sigma = 0.3$, $y_0 = 0.04$

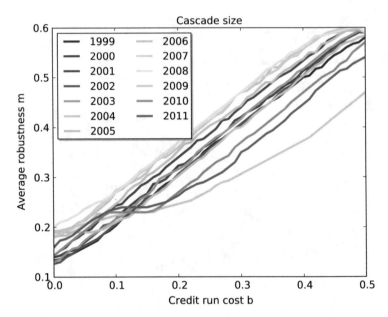

Fig. 12.8 Frontier of large cascades evolution of the e MID market in the period between January 1999 and January 2011 under random exogenous defaults and random individual robustness distribution. Impact of illiquidity on the structure of January of each year, $\sigma = 0.3$, $y_0 = 0.04$, $\gamma_s = 0.13$. For convenience, we use $\gamma_s = \gamma 10^{-3}$

as the amplification phenomena described in the model are less accentuated in this case. Quite naturally, since the ECB is not present in this dataset (but could be regarded as an external node), the model does not capture the level of risk transferred to the ECB itself. The external node would appear as a *node of last resort* to avoid systemic collapses.

Interestingly, these results can be thought as an interesting confirmation of the usefulness of these models in assessing the systemic impact of a shock from a macro-prudential point of view, by considering different network types for different levels of capital ratios and illiquidity on the market. Therefore, adopting this approach can be useful from a policy and regulatory perspective in that supervisors could design and implement more sophisticated capital requirements *ex-ante* and ad hoc liquidity scheme provisioning *ex-post*.

12.4.2 DebtRank

Introduced in Ref. [9], DebtRank measures the fraction of the total economic value that is affected by the default or distress of a financial institution. Recently, a new stress-test framework based on DebtRank has been proposed in Ref. [11], where the two networks of leverage presented in Sect. 12.2 represent the key concepts to understand the dynamics of distress propagation. Consider the network of financial interdependencies where the element A_{ij} represents the amount of assets i invests in the funding of j. The marginal column sums of the weighted adjacency matrix $A_i = \sum_j A_{ij}$ represents the total assets of i invested in funding within the network. Each node i is also endowed with a capital E_i and defaults when $E_i \leq 0$. The capital acts as buffer for i against shocks (see Refs. [19] and [33]) and can be thought, from a regulatory perspective, as the so-called Tier 1.

In this framework, liabilities are at face value whereas assets are marked to market. The main idea behind DebtRank is to capture distress induced not only by events of default, but also by a decrease in equity value due to a shock affecting the node, regardless of whether the node defaults or not. If the value of the equity of a node i decreases, this will decrease its *distance to default* as bank i will be less likely to repay its obligations in case of further distress. As a consequence, the market value of i's liabilities will decline and, as i's liabilities are on the asset side of the balance sheets of other institutions, the distress will propagate onto the equity value of neighbors, neighbors of neighbors, and so on.

However, the amount by which market values should decrease poses a non-trivial problem to solve, as it is determined by the probability of default of i and the recovery rate on assets held by i, both depending, in turn, by their market value. To our knowledge, there is no model in the literature dealing with this problem within a network system.

A reasonable conjecture would be that of a non-linear relation between losses on equity and losses on liabilities. In fact, on the one hand, it seems reasonable to assume that limited losses on equity may have little to none effect on the value of the liabilities. On the other hand, higher losses would reflect almost entirely and, especially when losses complete deplete the capital and therefore the equity value, this should in principle lead to the same impact on all liabilities.

Since neither the theory so far developed allows to capture this level of complexity nor the empirical evidence has evidenced particular significant patterns, we assume a simple linear impact, captured by the previously defined matrix A_{ij}. In other words, if E_i decreases by a proportion p, so does A_{ji}, $\forall j$ and each j faces a loss of pA_{ji}. It is immediate to show that, in case of default of i, then the loss for j is the entire amount due by i, $1 * A_{ji}$. As a result of this, also the equity of each j that has invested in i is impacted by the loss and its equity E_j will decrease in price as well. This distress mechanism propagates along all counterparties in the network. The linearity assumption must be intended as a paradigmatic case and future work will deal with more refined adjustment schemes.

We are therefore interested in the dynamics of the equity levels for each institutions. The DebtRank dynamics can be indeed thought as a process, starting at 0 and ending at T, where no further impact is seen. We can then see the evolutions of the equity levels $E_i(t)$, $t = 0, 1, \ldots, T$. Let

$$h_i(t) = \frac{E_i(0) - E_i(t)}{E_i(0)} = 1 - \frac{E_i(t)}{E_i(0)} \tag{12.4}$$

be the *cumulative relative loss on equity* from 0 to t of the i-th institution. $h_i(t, 0) \in [0, 1]$ and the default event of i at t implies $h_i(t, 0) = 1$.

The further propagation along the network is the key aspect of DebtRank. The mechanism proposed is based on feedback centrality, where the impact of the distress of an initial node across the whole network is computed recursively. However, unlike previously defined models of distress based on default cascade dynamics, DebtRank differs in that the distress propagates along the network even below the threshold of default. In order to quantify the indirect impact of i on all its successors (i.e. all the nodes that can be reached from i), a process is introduced: as a the value of the liabilities of a debtor decline in value, each credit will reduce their equity in order to face these losses. The dynamic reads:

$$h_i(t) = \min \left\{ 1, h_i(t-1) + \sum_{j \in S_A} \frac{A_{ij}(t)}{E_i} h_j(t-1) \right\} =$$

$$= \min \left\{ 1, h_i(t-1) + \sum_{j \in S_A} l_{ij}(t) h_j(t-1) \right\} \qquad (12.5)$$

This iteration goes on for a finite number of steps T, at which point all nodes will be either defaulted or still active. The recursive process comes into play as a node becomes distressed at t when a predecessor went into distress and at least one of its predecessors was in distress at $t-1$. Defaulted nodes remove their links from the network and, formally, they can be considered isolated nodes. This is formally obtained by summing over the element in set S_A in Eq. (12.5).[5]

In Ref. [11], the above described distress dynamics, starts ($t = 1$) because of an initial shock r on the unit value of the external assets of each bank. This results in the so-called "first-round" effects and banks record the relative equity loss $h_i(1) = l_i^e r$. Once this process starts, at $t = 2, 3, \ldots, T$, the reverberation dynamics follows Eq. (12.5), leading to "second round effects". In Ref. [11], $h_i(t)$ is referred to as the *individual vulnerability* of each bank i. This allows to compute loss distributions under different scenarios and/or distribution of the initial shock r. In particular, it is possible to compute also a *global vulnerability*, by considering the weighted average of each individual vulnerability, with weights given by the relative equity at time $t = 0$:

$$H(t) = \sum_i \frac{E_i(0)}{\sum_j E_j(0)} h_i(t)$$

---

[5]Further technical details on the motivations behind this dynamics, including a formal derivation, can be found in Ref. [11].

which expresses the relative amount of equity lost at time t. Different scenarios for the initial shock r can be adopted, allowing to compute loss distributions and standard risk measures for the individual institution and for the system as a whole. Further details on the stress-test framework are covered in Sect. 12.6.

To conclude the discussion on the DebtRank dynamics, we draw the readers' attention again to Eq. (12.5), which allows for an estimation of an upper bound of the overall network effect from the initial distress to the final effect. Locally (i.e., at each time step of the process), leverage plays the key role in that the impact is locally proportional to the leverage of the distressed exposures. A further characterization can be found as, indeed, the non-linear operator in Eq. (12.5) is bounded by the first eigenvalue of the leverage matrix conditioned to the sum of those institutions that have experienced a loss in equity for the first time. As such eigenvalue is increasing with respect to the overall levels of leverage, it is possible to link again the level of distress with the leverage. A formal derivation and further results are beyond the scope of the present work.

12.4.3 Prociclicality and Systemic Risk

In Ref. [40], the authors consider again a situation of interlocked balance sheets (Fig. 12.9), where the above mentioned leverage amplification is coupled with common asset exposures. In line with necessary adjustments to regulatory requirements, the balance sheets' dynamics is modeled in a dynamic stochastic settings, so that banks adjust their Value-at-Risk to specific target levels. The idea of *procyclicality* is explored within a framework where banks must be compliant to market risk-based capital requirements and asset market liquidity. Banks' compliance and liquidity in the assets constitute the two dimensions of a *table of market procyclicality* (Fig. 12.10). By assuming common asset-price shocks, a probability surface of

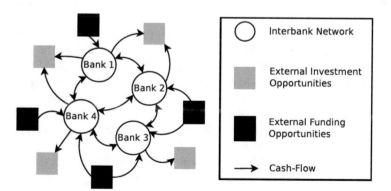

Fig. 12.9 The financial system as in Ref. [40]. The *nexus* hereby reflects the balance sheet structure of banks as in Fig. 12.2

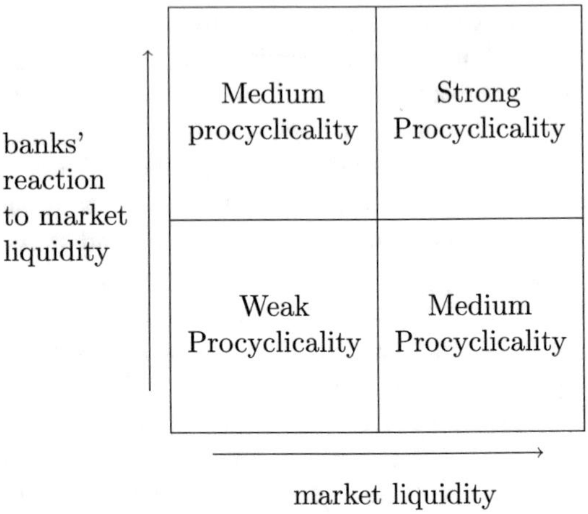

Fig. 12.10 The
procyclicality table

Medium procyclicality	Strong Procyclicality
Weak Procyclicality	Medium Procyclicality

banks' reaction to market liquidity

market liquidity

systemic default is derived. In particular, the system is perturbed with an aggregate asset price shock which is endogenously determined within banking activity, thus inducing a self-fulfilling dynamics. In particular, banks with high leverage will tend to liquidate negatively shocked assets, thus leading to a potential devaluation spiral. Again, overlapping portfolios, together with interbank claims generate prices drops. An inverse relationship between the economic cycle, market liquidity and capital requirement is found: when markets are liquid, strong capital requirements (which are usually thought as being procyclical) do not increase default probabilities; on the contrary, when the market is illiquid, weak capital requirements correspond to higher probability of default.

The main findings show that systemic risk keeps being high if three conditions occur at the same time: (1) bank compliance to target measure is strong, (2) the market is illiquid and (3) the structure is imbalanced towards external funds.

Even though banks might be perfectly rational from an individual point of view, from a systemic perspective this results in unwanted results. In fact, when banks are synchronized, small effect are amplified.

From an empirical perspective, the U.S. sub-prime crisis (2007–2009) can be taken as a paradigmatic example. Assets in the model would be the mortgage-backed securities which, after the shock suffered the blow up of the U.S. housing bubble. Being marked-to-marked and highly interconnected (via interbank lending), many large institutions stopped lending to each other, freezing the interbank market. Consistently with the findings of the mode, the Federal Reserve did indeed inject additional liquidity and bought part of the mortgage-backed securities, thus reducing banks' exposures towards that class of assets.

This has macroprudential implications as risk management needs to be accompanied by ad hoc monetary policies aimed at compensating market liquidity in presence of aggregated shocks.

12.5   Network of Financial Networks: Selected Empirical Works

12.5.1   The Interbank Network as a Multiplex

From the perspective adopted in the paper, another aspect needs to be taken into account: modeling the interbank network as a one-layered network is a limiting assumption. In fact, banks are increasingly connected along different dimensions. Though useful as a first analysis, considering only one single layer captures spillovers over the financial network as a whole, ignoring that distress can propagate along different dimensions. Introducing more layers in the network allows for a richer framework, where more types of relationships may occur. In order to capture systemic distress for banks, in particular, it is useful to identify two dimensions in the banking multiple network: the *type* of contract and the *maturity* of the contract.

Limiting the analysis to only one layer can indeed lead to over simplistic results. For example, a shock on a short-term loan could lead to a liquidity distress that can reverberate onto the other layers and generate, e.g., counterparty risk (in the long-term linkage). A multi-layered network modeling framework is more apt at capturing the increasingly complex and heterogeneous nature of financial linkages. Figure 12.11 shows a stylized representation of the multi-layer interbank network.

The multi-layered structure of the interbank market is modeled, for example, in Ref. [34] by developing an agent-based model. They focus on three different layers: long term loans (thus capturing direct and indirect counterparty risk), short term loans (direct and indirect liquidity risk) and common exposure to a certain asset class (direct and indirect common exposure). Multiple network relationships amongst banks can go beyond mere balance sheet exposures [13].

In Ref. [37] a descriptive analysis of the joint evolution of the interbank network for credit and the interbank network for Credit Default Swaps (CDSs), i.e. two interbank over-the-counter (OTC) markets for banking liquidity. Thus, the network considered can be seen as a multiplex network with two different link types. The empirical analysis focuses specifically on the German interbank network exploiting data for more than 2000 German banks from the German large credit register. The data span from the first quarter of 2002 to the third quarter of 2012, thus encompassing different historical periods, including the financial crisis.

As a matter of fact, OTC markets are non-transparent due to the lack of data availability and therefore can be a particular source of systemic risk. By coupling the two networks in a multiplex, the authors can investigate potential correlations in the behavior of banks along the two networks. The paper finds a striking level of stability for most network variable over the entire period, thus showing that interbank markets can be much less variable than expected. A certain variability in the volumes concentration for the CDS was found, in that it increased remarkably between 2005 and 2008 (i.e. in the build-up of the crisis) and decreased for the following 3 years. Another interesting results it that the intermediation chains remained constant in time, despite periods of high market volatility. Last, the data

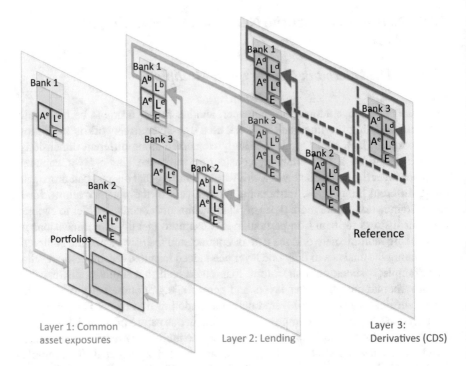

Fig. 12.11 *The interbank network as a multiple network.* A schematic representation of the balance sheets of three banks akin to that of Fig. 12.2 is reported. The first layer (*light blue color*, on the *left*) represents the exposures banks have to common assets. Shocks on this part of the nexus will impair the asset side of the banks forcing for a balance sheet reduction. The layer in the center (*light orange*) represents the "traditional" interbank lending. The last layer on the right (*light red*) represents the CDS network, where pairs of banks are connected if they engage in one of these contracts (links in *red*). Interestingly, other banks can be the reference entity for a contract (links in *purple*)

shows a large degree of correlation between the network networks, thus implying a certain level of similarity between the two layers of the multiplex.

Although interbank markets (including the German case) have been previously extensively studied, this contribution represents the first where the CDS network has been analyzed relying on actual exposure rather than reconstructed or inferred networks (Fig. 12.12).

12.5.2   CDS Networks

Another level on the multiplex network of the interbank exposures can be found in the subscription, by two different banking institutions, of a Credit Default Swap (CDS), a particular type of derivative contract. A CDS can be briefly described as

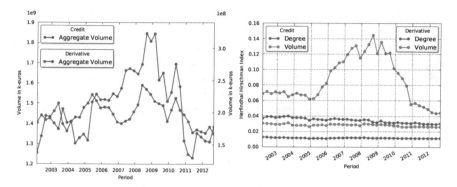

Fig. 12.12 From Ref. [37]. The graph on the *left* shows the dynamics (2002–2012) for the total aggregate volume in the credit (*red line, left axis*) and the CDS (*blue line, right axis*). The graph on the *right* shows the dynamics of the Herfindahl index for the degree and volume for the credit network (*red lines*) and the CDS (*blue lines*)

follows: two parties i and j agree on the payment of a sum (e.g. from i to j) in the case of the default of a so-called third party underlying or reference entity k. The exposure of one banking institution towards, say, i to j can be big enough to trigger the default of i. In this case, if i is the reference entity of other CDSs in the banking system, new defaults can be triggered, therefore generating a cascade.

CDS data are often confidential and the contracts themselves traded over the counter. As such, it is difficult to obtain data on CDSs that would involve some sort of exposure for all (or at least a relevant subset of) banking institutions. However, it is still possible to infer some sort of dependence structure starting from the time series of the spreads. In ideal market conditions, in fact, spreads on Credit Default Swaps should fully reflect the risk of the default of the underlying entity. Despite this assumption, it has now been widely recognized that CDS spreads have followed, rather than anticipated the financial breakdown. If CDSs were to price correctly the risk of defaults, then the network of their correlations could be used an early warning system, should it show some form of structural breakdown before a collapse.

Following this research question, in Ref. [35], the authors analyzed 176 CDS time series of different financial institutions (for a time period spanning from 2002 to 2011), by building different networks of correlation. They observed that these networks show some type of structural change at the *onset* of the crisis and not before. Assuming that these network represent a proxy of the dependence structure between pairs of companies, they further build a stress-test exercise based on Group DebtRank. Interestingly, they found that systemic risk prior to the crisis (before 2008) rises only when taking into account a specific macroeconomic indicator reflecting variations in the price of assets related to exposures in the housing sector. By considering this macroeconomic proxy, the paper provides an argument in favor of the incorporation of these quantities in order to detect systemic distress.

Methods: Data and Network Construction

The HPI is defined as the ratio of the house price index over the yearly household disposable income (source: http://www.fhfa.gov) and therefore expresses house prices in terms of income. CDS data consists in 176 daily time series with maturity 5 years of top US and European financial institutions. Networks are built in six different ways and all stem from a measure of comovements for the time series: (i) a correlation network:

$$w_{ij} = 1 - \frac{\sqrt{2(1 - \rho_{ij})}}{2}, \tag{12.6}$$

where ρ_{ij} is the Pearson correlation coefficient (statistically significant correlations are then considered); (ii) delayed correlation; (iii) Non-linear correlation; (iv) Granger causality; (v) drawups.

Main Findings

A simple visual analysis of the daily time series for the CDS spread (Fig. 12.13), show that the period 2002–2007 is characterized by low spread values and low volatility (including some minor bursts in 2005). From the beginning of 2007, spreads rapidly increase reaching values sometimes 10 or more times larger than they were before. The same figure shows also the behavior of the average market capitalization for the same set of companies. After a steady increase in the period 2002–2006, from the beginning of 2007 the average value decreases steadily and

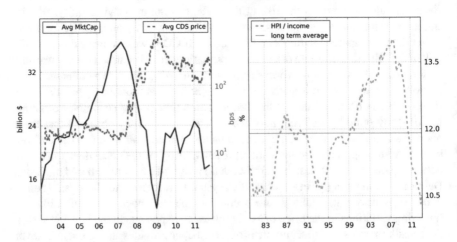

Fig. 12.13 The *left graph* shows the average CDS price for the 176 institutions (in *red*) and the average market capitalization (*black*). The graph on the *right* shows the house price index divided by households' disposable income

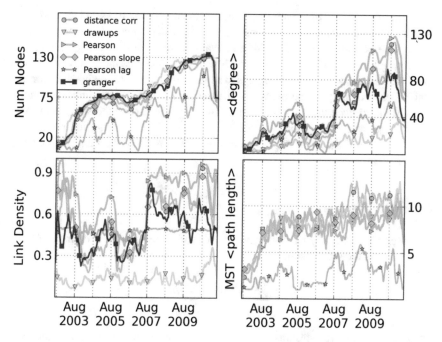

Fig. 12.14 Time-series of the main network statistics with the different network construction methods. Number of nodes (*top left*), degree (*top right*), density (*bottom left*). Average path length of the minimum spanning tree (*bottom right*)

by the end of 2008 drops by almost 50 %. The point of inversion for the two curves occurs almost at the same time. For the sake of our successive analyses, in Fig. 12.13 (right), the ratio of the house price index over the household disposable income, denoted as HPI is compared.

Network Measures

The *size N* of the network is the number of non-isolated nodes. From a time series perspective, this implies that the times series of one institution has to comove with at least one other time series in a statistically significant way. The *density* of each network is defined as the ratio between the existing links and the number of all possible links. In other words, it measures the fraction of time series that have statistically significant comovements. Figure 12.14 (bottom left graph), shows a generally high network density, ranging from 20 % to 50 % depending on the method and time periods. Interestingly, at the onset of the crisis, the density increases. In other words, the level of interdependence between institutions in the network increases, thus carrying potential implication for enhanced distress propagation.

A gradual increase of the number of nodes, the number of links and their weights is observable in time. It is also interesting to observe the structural change occurring in the time period 2005–2007, with a relevant decrease in the average degree and density. Notice that, with the rolling window method, the time interval is backward–looking. The above described findings are quite similar across methods. Given the high level of density of the networks, the authors extract the Minimum Spanning Tree (MST) of each undirected network [14]. A particularly interesting quantity is the average path among nodes in the MST. Figure 12.14 shows (bottom right) a sharp increase of the average path within the MST in 2003, with a slight increase later on.

Group DebtRank

Consider the subset S_f of nodes in the network, then one can define a *Group DebtRank*, as the simple summation over all nodes belonging to S_f:

$$R = \sum_{j \in S_f} h_j(T)v_j - \sum_{j \in S_f} h_j(1)v_j$$

or, in other terms, R measures the distress *induced* by the system, not taking into account the initial (exogenous) distress, hence giving a useful proxy for the network effects.

In the two graphs reported in Fig. 12.15, the dynamics of Group DebtRank has been reported for two different network construction methods (the Pearson on the

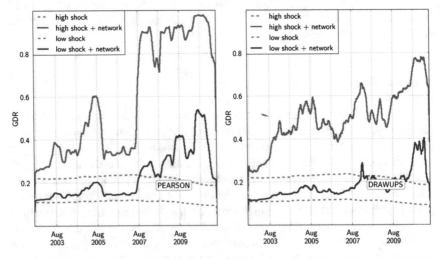

Fig. 12.15 Values of Group DebtRank for the Pearson network method (*left graph*) and the Drawups method (*right graph*) for different scenarios

left and the drawups method on the right). Higher systemic risk is reflected by higher Group DebtRank values. Solid lines represents the dynamics of Group DebtRank resulting from the combination of network effects and the macroeconomic shock, whereas dashed lines refer to the impact on the system if only the shock is considered. The colors refer to two different scenarios, a higher shock scenario (red lines) and a lower shock scenario (blue lines). Both graphs show the impact of network effects in the build up of the crisis. The distance between lines of the same color can be interpreted as the amplification effect of the network. As shown in the picture, such amplifications can even reach a factor of 4.

CDS Networks: Conclusions

From a pricing perspective, findings show that networks built from time series do not capture systemic risk build-up in the system and therefore suggest that CDS do not price correctly the risk of default. Comovement networks do not suggest particular additional information. More to this, adopting a modified Group DebtRank approach with macro-indicators leads to more interesting results that confirm the link towards another part of the nexus, i.e. common exposures to assets.

12.6   Common Asset Exposures: A Novel Approach Based on DebtRank

From our "nexus of interconnected networks" approach, the mechanics of common exposures in determining fire sales can be further formalized in network terms. In line with the balance sheet representation of Fig. 12.2 we introduce three networks:

1. the already examined monopartite $(N \times N)$ interbank network, whose weighted adjacency matrix is represented by A_{ij}^b;
2. the bipartite $(N \times M)$ bank/asset network, with weighted adjacency matrix A_{ik}^e;
3. the monopartite $(M \times M)$ asset network.

The asset network represents the *common asset dynamics*. It goes beyond the scope of this work to explain the type of commonality each pair of assets shares. However, from an intuitive point of view, these relationships can be thought of as common price co-movements or, of a more mechanical nature, such as the relation occurring between a derivative and its underlying. The network of common exposure is then a key aspect of the distress process. Work in this direction has been done, for example, in Ref. [31].

12.6.1 Network of Assets

A detailed explanation of the network of assets goes beyond the scope of this work. However, we would like to draw the reader's attention towards the three main points.

1. Assets do form a network on their own, by being linked because of various economic reasons. For example, it is a reasonable assumption that the equity price of a company active in the construction sector of a particular country would somehow depend on the housing investment in that specific country. These effects can be also generated by the productive structure of an economy.
2. Common exposures are not only related to production but also to a signaling effect of market prices. For example if two firms have a similar asset structure, bad news about the asset structure of one of the two firms will reflect onto the other. So common exposures matter in determining the structure of the asset network. Therefore, mapping common exposures and asset overlap becomes a crucial aspect represent with a bipartite network on its own.
3. A more mechanicistic level exists between a derivative and its underlying. This is, however, a quite complex notion that would require extensive research.

12.6.2 Estimations Using the DebtRank Approach

Common exposures towards the same asset class represent another level of inter-connectedness between financial institutions, therefore they can be aptly integrated in the nexus of networks we explore in this Chapter. In order to do so, we discuss the main theoretical and empirical results of the stress-test framework proposed in Ref. [11]. The authors focus:

1. from the theoretical point of view, on the development of a three-round proce-dure, that models the dynamics of equity loss at the first (initial shock on external assets), second (reverberation on the interbank market) and third (fire sales);
2. from the empirical point of view, they apply the stress-test framework to a set of 183 traded European banks.

The theoretical contribution builds on the leverage decomposition analyzed in Sect. 12.2 and on the DebtRank distress propagation dynamics illustrated in Sect. 12.4.2.

Consistently with Sect. 12.2, let $l_e = \frac{A_i^e}{E_i}$ and $l_b = \frac{A_i^b}{E_i}$ be respectively the leverage w.r.t. the external assets and the leverage w.r.t. the interbank lending market. If banks have *common asset exposures*, a shock on the external assets would hit a number of institution within the financial system. For sake of simplicity, the authors assume that banks have common exposure to *only one* asset class. This, although being a restrictive assumption, can be interpreted as having a number m of assets with unitary correlation, so that the impact of the price is uniform and can be helpful in providing an upper bound for the computation of the total equity loss. The total value of the asset is given initially by $A_i^E = Q_i p(0)$, where Q_i is the quantity (e.g. shares) of asset held by bank i and $p(0)$ is the initial market price of the asset.

A Three-Round Stress-Test

Now, suppose that the price of the asset drops by an amount $r < 0$. The equity loss dynamics is captured by Eq. (12.4). The following dynamic will take place:

1. **First round**. At the *first round*, every bank j holding the asset will experience a loss in equity equal to $l_j^e r$: therefore the impact is simply proportional to the leverage ratio.
2. **Second round**. At the *second round*, the loss in equity will propagate towards the neighbors (lenders) of j in the interbank network. If another bank i has lent to bank j, the value of these obligations will be marked-to-market downwards. We assume, consistently with the DebtRank approach, such reduction in the value of the interbank assets to be proportional to the loss of value of the equity of the neighbors (Eq. (12.5)). Assuming no default, all banks i exposed to j will therefore lose value in equity equal to:

$$h_i(2) = h_i(1) + \sum_j l_{ij} l_j^e r$$

   In other terms, the impact is amplified *multiplicatively* by the product of the two leverage ratios. The compounding effects of the two leverage subcomponents is made explicit by computing a *first order approximation* (details on the derivation in Ref. [11]):

$$h_i(2) = l_i^e r + l_i^b l_i^e r = r\, l_i^e (1 + l_i^b)$$

3. **Third round**. At the *third round* banks attempt to restore the previous leverage ratio before the shock. If this is not possible by raising more equity, banks will start selling assets in order to reduce their asset size and therefore meeting leverage target [2, 40]. Each bank will try to target the original leverage levels $l_i(0)$, by selling external assets in exchange of cash. The large aggregate supply of asset will have a strong impact in reducing asset prices and therefore banks will experience further losses (*third round* effects).

   Since it goes beyond the scope of this Chapter to provide an analytical derivation of the third round effects, we will hereby sketch the main points (the reader can refer to [11] for the complete derivation). First, the loss on equity before the third round is given by $h_i(T) = l_i^e(0)r(1) + \sum_j l_{ij}^b l_j^e s(1)$. After some passages, one can obtain a formula describing the quantity of external assets sold by each bank as:

$$\frac{\Delta Q_i}{Q_i(0)} = \frac{D_i(0)}{Q_i \hat{p}} \left(r(1) + \sum_j l_{ij}^b l_j^e \right) \tag{12.7}$$

By assuming a simple linear impact on prices, with η being the coefficient used to measure the impact on price of the quantity of external assets sold, the authors obtain the following analytical expression for the third round effects:

$$h_i = \left(l_i^e + l_i^e l_i^e \right) r + \eta \frac{D_i}{Q_i} \left(l_i^e + l_i^b l_i^e \right)^2 r \tag{12.8}$$

which shows how the original shock compounds with the square of the product of the two leverage subcomponents. Empirically, these quantities are typically larger than one, therefore the overall effect of a shock can be extremely large. In this light, Eq. (12.8) shows that:

a. the two leverage components (referring to two different networks in the nexus) actually have a compounding effect;
b. neglecting the interaction of the two networks would lead to an underestimation of the total possible effects.

The authors also provide an empirical application on a set of 183 listed European banks. In Fig. 12.16, results of this exercise are reported. The left panel shows a decomposition of first, second and third round effects, for an initial shock of $r = 0.01$ on the *global equity loss* (global vulnerability). We observe that second

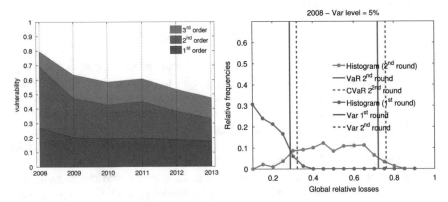

Fig. 12.16 Global vulnerability (From Ref. [11]). (*Left*) Time evolution for the global vulnerability over a sample of 183 European banks, for an initial shock on external assets of 1 %. The breakdown in the stacked graph represents first, second and third order effects. (*Right*) Loss distribution, VaR and CVaR at the first and second round, initial shocks on external assets are drawn from a Beta distribution (see Ref. [11] for the technical details)

and third round are at least of the same order of magnitude of the first round effects, providing a compelling argument towards the inclusion of network and common asset exposures effects in the current stress test frameworks. The right panel of Fig. 12.16 shows how, by generating random value for r from a Beta distribution, the framework allows for the estimation of loss distributions and for the computation of standard risk measures such as VaR and CVaR. It also shows how these risk measures can be severely underestimated if network effects are neglected.

12.7   Conclusions and Future Research

In this Chapter, we explored the idea that the financial system ought to be seen as a *nexus* of interconnected networks, rather than an isolated networked system. Such interconnections stem from the balance sheet structure of banks and they are quite heterogeneous in nature.

By seeing the balance sheet model in terms of leverage, we have provided a formulation of DebtRank that shows how leverage compounds in the final effects in terms of distress. We then explored a network of exposures towards other macroeconomic sectors, illustrating the concepts of network effects *within* a macroeconomic block and network effects *within* blocks. By reviewing a series of recent theoretical and empirical works, we show the importance of understanding these further level of interconnections.

Particular emphasis has been given to networks of common exposures, in that they can exacerbate initial shocks and leads to spirals of asset devaluation. In this light, we sketched some modeling ideas to extend DebtRank in order to include common exposures.

Further research will deal with refinement of the leverage-based DebtRank approach. In particular, attention should be devoted to the understanding of the relationship between individual and systemic risk from a leverage perspective. From a regulatory point of view, this notion will then imply a reflection on the mutual relationship between the level of interconnection of an institution and capital requirements.

References

1. Acharya, V.V.: A theory of systemic risk and design of prudential bank regulation. J. Financ. Stab. **5**(3), 224–255 (2009)
2. Adrian, T., Shin, H.S.: Liquidity and leverage. J. Financ. Intermed. **19**(3), 418–437 (2010)
3. Allen, F., Gale, D.: Financial contagion. J. Pol. Econ. **108**(1), 1–33 (2000)
4. Allen, F., Babus, A., Carletti, E.: Asset commonality, debt maturity and systemic risk. J. Financ. Econ. **104**(3), 519–534 (2012)
5. Bakk-Simon, K., Borgioli, S., Girón, C., Hempell, H.S., Maddaloni, A., Recine, F., Rosati, S.: Shadow banking in the euro area: an overview. ECB Occasional Paper, No. 133 (2011)

6. Battiston, S., Caldarelli, G.: Systemic risk in financial networks. J. Financ. Manage. Mark. Inst. **2**, 129–154 (2013)
7. Battiston, S., Delli Gatti, D., Gallegati, M., Greenwald, B.C.N., Stiglitz, J.E.: Credit default cascades: when does risk diversification increase stability? J. Financ. Stab. **8**(3), 138–149 (2012a)
8. Battiston, S., Gatti, D.D., Gallegati, M., Greenwald, B.C.N., Stiglitz, J.E.: Liaisons dangereuses: increasing connectivity, risk sharing, and systemic risk. J. Econ. Dyn. Control **36**(8), 1121–1141 (2012b)
9. Battiston, S., Puliga, M., Kaushik, R., Tasca, P., Caldarelli, G.: DebtRank: too central to fail? Financial networks, the FED and systemic risk. Sci. Rep. **2**, 541 (2012c)
10. Battiston, S., Caldarelli, G., Georg, C.-P., May, R., Stiglitz, J.: Complex derivatives. Nat. Phys. **9**(3), 123–125 (2013)
11. Battiston, S., Caldarelli, G., D'Errico, M., Gurciullo, S.: Leveraging the network: a stress-test framework based on debtrank. Stat. Risk Model. (2015). SSRN eLibrary
12. Beale, N., Rand, D.G., Battey, H., Croxson, K., May, R.M., Nowak, M.A.: Individual versus systemic risk and the Regulator's Dilemma. Proc. Natl. Acad. Sci. **108**(31), 12647–12652 (2011)
13. Bonacina, F., D'Errico, M., Moretto, E., Stefani, S., Torriero, A., Zambruno, G.: A multiple network approach to corporate governance. Qual. Quant. **49**(4), 1585–1595 (2015)
14. Bonanno, G., Caldarelli, G., Lillo, F., Mantegna, R.N.: Topology of correlation-based minimal spanning trees in real and model markets. Phys. Rev. E **68**(4), 46130 (2003)
15. Boss, M., Elsinger, H., Summer, M., Thurner, S.: An empirical analysis of the network structure of the Austrian interbank market. Oesterreichesche Natl. Financ. Stab. Rep. **7**, 77–87 (2004)
16. Burkholz, R., Garas, A., Schweitzer, F.: How damage diversification can reduce systemic risk (2015). arXiv:1503.00925
17. Cajueiro, D.O., Tabak, B.M.: The role of banks in the Brazilian interbank market: does bank type matter? Physica A: Stat. Mech. Appl. **387**(27), 6825–6836 (2008)
18. Castrén, O., Rancan, M.: Macro-networks: an application to euro area financial accounts. J. Bank. Finance **46**, 43–58 (2014)
19. Cont, R., Moussa, A., Santos, E.B.: Network structure and systemic risk in banking systems. SSRN eLibrary (2010)
20. Craig, B., Von Peter, G.: Interbank tiering and money center banks. Available SSRN 1687281, (10–14) (2010)
21. de Masi, G., Iori, G., Caldarelli, G.: Fitness model for the Italian interbank money market. Phys. Rev. E **74**(6) (2006)
22. Doyle, J.C., Alderson, D.L., Li, L., Low, S., Roughan, M., Shalunov, S., Tanaka, R., Willinger, W.: The "Robust yet fragile" nature of the internet. Proc. Natl. Acad. Sci. USA **102**(41), 14497–14502 (2005)
23. Eisenberg, L., Noe, T.H.: Systemic risk in financial systems. Manage. Sci. **47**(2), 236–249 (2001)
24. Elsinger, H., Lehar, A., Summer, M.: Risk assessment for banking systems. Manage. Sci. **52**(9), 1301–1314 (2006)
25. European Commission: Financing the real economy. Product Market Review 2013. Technical report (2013)
26. Gai, P., Kapadia, S.: Contagion in financial networks. Proc. R. Soc. A Math. Phys. Eng. Sci. **466**(2120), 2401–2423 (2010)
27. Gai, P., Haldane, A., Kapadia, S.: Complexity, concentration and contagion. J. Monet. Econ. **58**(5), 453–470 (2011)
28. Greenwald, B.C.N., Stiglitz, J.E.: Financial market imperfections and business cycles. Q. J. Econ. **108**, 77–114 (1993)
29. Haldane, A.G.: Rethinking the Financial Network. Speech Delivered to the Financial Student Association, Amsterdam (2009)
30. Haldane, A.G., May, R.M.: Systemic risk in banking ecosystems. Nature **469**(7330), 351–355 (2011)

31. Huang, X., Vodenska, I., Havlin, S., Stanley, H.E.: Cascading failures in bi-partite graphs: model for systemic risk propagation. Sci. Rep. **3**, 1219 (2013)
32. Iori, G., De Masi, G., Precup, O.V., Gabbi, G., Caldarelli, G.: A network analysis of the Italian overnight money market. J. Econ. Dyn. Control **32**(1), 259–278 (2008)
33. Mistrulli, P.E.: Assessing financial contagion in the interbank market: maximum entropy versus observed interbank lending patterns. J. Bank. Finance **35**(5), 1114–1127 (2011)
34. Montagna, M., Kok, C.: Multi-layered interbank model for assessing systemic risk. Technical report, Kiel Working Paper (2013)
35. Puliga, M., Caldarelli, G., Battiston, S.: Credit default swaps networks and systemic risk. Sci. Rep. **4** (2014)
36. Roukny, T., Bersini, H., Pirotte, H., Caldarelli, G., Battiston, S.: Default cascades in complex networks: topology and systemic risk. Sci. Rep. **3**, 2759 (2013)
37. Roukny, T., George, C.-P., Battiston, S.: A network analysis of the evolution of the German interbank market. Dtsch. Bundesbank Discuss. Pap. 22/2014 (2014)
38. Soramäki, K., Bech, M.L., Arnold, J., Glass, R.J., Beyeler, W.E.: The topology of interbank payment flows. Physica A: Stat. Mech. Appl. **379**(1), 317–333 (2007)
39. Stiglitz, J.E.: Risk and global economic architecture: why full financial integration may be undesirable. Am. Econ. Rev. **100**(2), 388–392 (2010)
40. Tasca, P., Battiston, S.: Market procyclicality and systemic risk. MPRA Pap. No. 45156 (2013)
41. Upper, C., Worms, A.: Estimating bilateral exposures in the German interbank market: is there a danger of contagion? Eur. Econ. Rev. **48**(4), 827–849 (2004)
42. Vitali, S., Glattfelder, J.B., Battiston, S.: The network of global corporate control. PLoS ONE **6**(10), e25995 (2011). doi:10.1371/journal.pone.0025995

Printed in the United States
By Bookmasters